Cases on E-Readiness and Information Systems Management in Organizations:

Tools for Maximizing Strategic Alignment

Mustafa Alshawi
University of Salford, UK

Mohammed Arif
University of Salford, UK

Managing Director:	Lindsay Johnston
Senior Editorial Director:	Heather Probst
Book Production Manager:	Sean Woznicki
Development Manager:	Joel Gamon
Development Editor:	Joel Gamon
Acquisitions Editor:	Erika Gallagher
Typesetter:	Lisandro Gonzalez
Print Coordinator:	Jamie Snavely
Cover Design:	Nick Newcomer

Published in the United States of America by
    Business Science Reference (an imprint of IGI Global)
    701 E. Chocolate Avenue
    Hershey PA 17033
    Tel: 717-533-8845
    Fax: 717-533-8661
    E-mail: cust@igi-global.com
    Web site: http://www.igi-global.com

Library of Congress Cataloging-in-Publication Data

Cases on e-readiness and information systems management in organizations: tools for maximizing strategic alignment / Mustafa Alshawi and Mohammed Arif, editors.
    p. cm.
  Includes bibliographical references and index.
  ISBN 978-1-61350-311-9 (hbk.) -- ISBN 978-1-61350-312-6 (ebook) -- ISBN 978-1-61350-313-3 (print & perpetual access) 1. Information technology--Management. 2. Technological innovations--Management. 3. Organizational effectiveness. I. Alshawi, Mustafa. II. Arif, Mohammed. III. Title.
  HD30.2.C3788 2012
  004.068'4--dc23
                     2011031963

British Cataloguing in Publication Data
A Cataloguing in Publication record for this book is available from the British Library.

Table of Contents

Section 2
Issues in IT/IS Readiness

Detailed Table of Contents

Section 1
Cases in IT/IS Readiness

This chapter explains the concept of an IT/IS readiness maturity model including particular requirements in terms of four domains, embracing nine attributes: IT infrastructure (top management perception, systems and communication), people (skills, roles and responsibility of IT staff, user involvement), process, and work environment (organization behaviour, IT department, leadership). Each of the attributes consists of 14 factors: top management perception (drivers, systems requirements definition), systems and communication (focus, network communication), skills (type of skills, capability building), roles and responsibility of IT staff (position of IT/IS heads, roles of IT staff), user involvement, process (practices), organizational behaviour (characteristics), IT policy (control of IT/IS activities), and leadership (communication, participation). The following section describes the concept of readiness and maturity, the resources used for element extraction/adoption and the description of the model.

Higher education sector is notorious for lagging behind the industrial sector in the application of IT/IS systems and infrastructure. This chapter presents the application of the IT/IS readiness model in a higher education organization. This organisation was established in 1967 and currently has about 2,500 staff and 18,000 students, of which, 3,000 are international students from all over the world. The organization comprises of 14 schools and 13 research institutes and offers programmes various fields, which include virtual reality, magnetic and optics, business, law, genetic algorithms, health-related studies, and building construction. In 1996, Academic Division (AD) identified the need to improve the management of the student database due to the increase of students and programs offered by the organization. AD also identified that the Legacy Student Information System (SIS) was unable to cope with the increasing demand of data administration. This case study presents the overview of issues encountered while assessing the e-readiness of the organisation after most of the systems went live. Post implementation, the system has been able to reduce the redundancies in processes and has been able to provide a more effective support to students and staff. However, still there are several issues and conflicts that need to be resolved, and a radical rethink of the processes supporting the IT system is needed to achieve any further efficiency.

Construction sector is unique in a way because more than 90% companies are Small and Medium Sized Enterprise (SME). This chapter presents a case study of a construction company with past and current projects valued up to £15 million, and has completed a number of construction projects both in private and the public sectors including housing, commerce, leisure, health, education, retail, et cetera. The company operates out of multiple locations and decided to improve the tendering process using a new IT system. On applying the IT readiness model it was found that almost all the attributes identified were not at lower level 1. The practice achieved maturity in three areas – skills and leadership are identified at level 5, and roles & responsibility were identified at the top of the level.

This chapter provides the IT readiness assessment for before and after scenarios of IT systems implementation in a construction consultancy company providing multi-disciplinary services for the construction industry throughout the United Kingdom. The services offered include building surveying, quantity surveying, project management, civil and structural engineering design, and mechanical and electrical engineering design, among others. On application of the maturity model it was found that the overall processes for managing information are improving since the introduction of the new IT system. Prior to the project, the development of IT/IS was driven to perform daily work tasks that required the company to run a business. The new systems has streamlined the organization-wide communication, which the previous system did not have the capability to do, and to reduce cost for document reproduction. The level of IT skills prior to the project was relatively low; the introduction of the new system has helped the company to increase their staff's IT skills.

Chapter 5
Yasser Al Saleh, University of Salford, UK
Eric Lou, University of Salford, UK

This chapter presents the case of a bank that was established in late 1973 by an initiative from the Government, as a joint venture between the Ministry of Finance, the Central Bank, all commercial banks registered in the country, insurance companies, and some large industrial investment firms. The chapter presents the before and after analysis of IT readiness in the bank. Through the analysis it was realised that the use of IS/IT in the Bank is well behind that normally expected from a bank, which is set up to promote the industrial sector of a major finance centre. The Bank management needs to recognise that technology is changing too fast for the non-specialist to keep up. It is therefore an essential part of a truly professional relationship between a business and its IS/IT staff that they help management to understand that business opportunities which would arise from IS/IT utilization in new areas such as electronic commerce. This vision does not exist in the Bank, at least not in those professionals whose position would give them the necessary influence. This means that the Bank's top management need is to consider IS/IT as a strategic tool to achieve a competitive edge in pursuing the Bank's goals.

Chapter 6
Yasser Al Saleh, University of Salford, UK
Mohammed Arif, University of Salford, UK

This case study revolves around a governmental public service institution, which receives public and government money that it invests. There were several challenges associated with the implementation of the IT system to improve public service. It was found that the organizations need, in the contract, to have the qualifications of the vendor's staff, and agree that prior approval for any change of staff or new recruitment would be agreed beforehand. This is because the vendor's staff had a high turnover. Experienced staff, which were agreed upon by the organization, were assigned to the project for a short time, only at the beginning of the project. The lack of positive relationships between different groups in the organization caused resistance to the required changes in structure and processes. Because key staff considered keeping knowledge and experience to themselves as a job security tool, they were not forthcoming in cooperating with the project team. This was complicated by the almost complete absence of systems' documentation, and the little documentation that did exist was obsolete or not comprehensive. The void of decisive leadership by top management allowed the conflicts between different entities in the organization to go on in an increasing mode until the end of the project, which had a negative effect on the project success. The new system design was not successful in resolving the ownership of the data within the organization. This was an issue that caused user resistance to the project.

Section 2
Issues in IT/IS Readiness

Chapter 7

    David Rehak, VSB – Technical University of Ostrava, Czech Republic
    Monika Grasseova, University of Defence, Czech Republic

The chapter is focused mainly on assessing the factors of the external environment in the area of security of information systems in the organization through SWOT analysis. At first the method is characterized from the viewpoint of its purpose and nature. The emphasis is laid on the principles of SWOT analysis, the possible use of methods and tools, and also the most common problems occurring during the implementation of the analysis. The recommended methodical procedure for the implementation of SWOT analysis is described in another part of the chapter with individual phases and particular activities, which are appropriate to be carried out within these phases. The main part of the chapter is focused on the ways of semi-quantitative assessment of threats to the area of information systems of the organization, while evaluating their risks, and the assessment of opportunities, while evaluating their benefits. Both cases include a detailed description of procedure leading to an objective outcome during the classification of identified threats and opportunities according to the set criteria.

    John Effah, University of Ghana Business School, Ghana
    Ben Light, University of Salford, UK

The purpose of this study is to understand a small e-support firm's response to the local e-readiness and the global e-business environment in a developing context. E-Support firms provide Web development and consultancy services to user organizations, assisting them in their uptake and maintenance of their Internet applications. Within the e-readiness research area, little is known about e-support firms, particularly in connection with their interaction with their local and the global e-business environment. As yet the emphasis on e-readiness studies has been at the national level. Nevertheless, the e-support sector is very significant in the successful adoption and diffusion of the Internet and related applications in any economy. It is thus important to understand how such firms relate to their e-business environments. That said, this study draws on the interpretive case study of a small e-support firm in Ghana, a developing context, to investigate the firm's response to the e-readiness level of the local and the global e-business environment. Findings show that the firm could employ resources from the global environment to address most of the infrastructural challenges posed by a relatively poor local e-readiness context. However, its attempt to transfer advanced e-business technologies from the global e-business environment to the local e-business context did not succeed. This chapter offers implications for practice and research concerning the notion of reconciling local and global e-business environments in the small e-support sector.

    Vian Ahmed, University of Salford, UK
    Aisha Abuelmaatti, University of Salford, UK

Collaborative environments have been evolving and effectively employed in large organisations and are believed to have high potential for Small and Medium Enterprises (SMEs). This chapter shares the findings of a case study that was conducted on twelve companies in order to assess the use of collaborative environments and their adaptation approaches through interviews with senior level managers and end-users. The need for such case studies has risen from an intensive literature review which revealed that SMEs are key players within the construction industry; however, there seems to be little evidence of their utilisation of IT for collaborative learning environments. Therefore, this calls for the necessity to developing an approach blending the right combination of factors which are believed to contribute towards the improvement and implementation of collaborative environments and may affect their success.

Chapter 10

Ayman Altameem, University of Bradford, UK
Mohamed Zairi, University of Bradford, UK

The value that information technology (IT) can bring to organizations is clear, and few will dispute its potential. However, the literature shows that worldwide, many of the organizations adopting IT fail to achieve the desired results and that sometimes, the cost of failure can far exceed the expected benefits. There are several reasons why IT projects fail to deliver. This study is an attempt to bridge the gap in the existing literature by exploring the critical factors that affect IT adoption through a comprehensive benchmarking analysis, using secondary cases. The IT adoption in 100 organizations indicated in the literature, were scrutinized in all the cases analyzed in order to arrive at the most critical factors affecting IT adoption, as well as their degree of criticality. The study identifies twenty-four critical factors that must be carefully considered in IT adoption to attain successful outcomes.

Chapter 11

Masoud Mohammadian, University of Canberra, Australia
Ric Jentzsch, University of Canberra, Australia

IT management processes have been growing as the development of modern IT systems has grown. These are often complex with multiple interdependencies that can make it very difficult for Chief Information Officers (CIOs) to comprehend and be aware of potential risks. These risks have the potential to translate into decision making inefficiencies for an organization. Risk analysis for decision making in the planning and monitoring of these systems can be a complex and demanding task. Intelligent decision making in IT management processes and systems are a crucial element of an organization's success and its competitive position in the marketplace. This chapter considers the implementation of Fuzzy Cognitive Maps (FCM) to provide facilities to capture and represent complex relationships in an IT management process model. By using FCMs, CIOs can regularly review and improve their IT management processes and provide greater improvement in development, monitoring and maintenance of those processes. CIOs can perform what-if analysis to better understand vulnerabilities of their designed system.

Preface

INTRODUCTION

Growth and competitiveness of organizations can only be sustained through the pursuance of knowledge and innovation. The latter has historically been driven by rapid developments in ICT where ICT-based innovations have important implications on business and socioeconomic development. This is mainly due to their role in introducing and diffusing the concepts of high quality at low cost, knowledge sharing, community development, and promotion of equality. Organizations have invested heavily in ICT over the past two decades. Figures have shown investments in ICT may exceed 50% of annual capital investment and the average ICT expenditure could amount to 5% of organizational revenue.

The key element of success is accepting change and a new way of working. Organizations need to adapt and explore concurrent changes in today's borderless economy, or risk losing out; while taking into account the importance of understanding technology adoption and diffusion issues. Whilst a 'technology push' approach may bring about 'first comer' advantages to the organization, implementing advanced IT applications to create sustainable competitive advantage can only be leveraged by improving business processes in line with management objectives using IT as an enabler. Data and information are shared and distributed digitally, formulating a cheaper and more effective way of communication. Electronic processes can generate huge new wealth and can transform the way businesses are conducted in unprecedented ways.

There is no doubt that there has been heavy investment in ICT. However, many case studies have shown that investments are made in an isolated fashion and are not delivering strategic advantage for organizations. Despite large spending, the failure rate of not achieving the intended business objectives is also increasing. Work by Lientz and Larson in 2004 found that 40% of ICT projects failed to deliver tangible benefits, and that less than 50% are completed on-time and within budget. Another study by Alrashid et al (2009) reported that 178% of systems run over cost and 230% run over time. Furthermore, some projects were abandoned, significantly redirected, or kept alive despite business integration failures. In some cases, projects were kept alive for the sake of keeping the project alive.

This led to a perceived dissolution of IT's strategic benefits, which seriously affected future IT investment. At the same time, the development and delivery of innovative IT-based services, and the ability to respond to the rapid evolution of markets, places a considerable premium on leadership and management skills (BERR, 2008). Here lies the challenge: on one hand, executives realize the importance of business-based IT solutions (which are normally expensive and difficult to implement) and are in a position to finance this, while on the other, they have witnessed significant failures and missed opportunities in previous investments.

In this respect, there is a need to understand the environment within which IT-based business improvement can be successfully achieved and then identify the gaps that need to be addressed. This will place executives in a much better position to predict the level of change and resources needed to develop the capability of their organizations, specifically:

- The creation of an innovative working environment which is focused on developing and sustaining a highly skilled and flexible workforce with the skills to continuously introduce improvements through better and more streamlined business processes enabled by advanced IT;
- The achievement of effective business processes and improvement focused on improving the organization's efficiency by directly integrating IT with the corporate, strategic, and operational goals to ensure IT resources are 'in line' with business imperatives.

STRATEGIC POSITIONING OF IT SURVEY

The University of Salford and Construct IT (Alshawi et al, 2008) have undertaken a survey which was specifically designed to assess the awareness and understanding of Construction Industry Executives on a) the strategic benefits that IT can bring about to their organizations and b) on the critical elements that lead to the realization of these benefits; IT skills, business process management/reengineering, and IT strategies and c) the drivers behind the decisions on IT investment; the impact of new technologies, e-readiness of organizations, and financial returns on investment. Each question had five options describing evolution through maturity levels referring to three scenarios; 1995 thinking, 2007 practice, and 2007 thinking. The survey received 109 responses from 80 different Contractor and Consultant organizations in the UK, where both Chief Executives and IT (or Innovation) Directors were targeted.

The results clearly show that construction organizations today are acknowledging the strategic nature and significant contribution of IT whereby IT systems are now considered at an organizational-wide scope, rather than an individual application. IT strategies are now slowly being integrated into organizational business strategies and the impact of IT technologies are recognized for delivering competitive advantage for the future. This is a significantly positive sign that the Construction Industry is moving forward towards utilization of IT; most importantly, with a similar set of thinking (contractors and consultants). On the other hand, construction organizations today have their IT investments influenced by the state of readiness of the organization to successfully receive new and future IT investments. The main findings are summarized as follows:

1. Although the industry is aware of the strategic benefits of IT, it has not yet been attained
2. Financial return on IT investment is still being practiced in spite of the industry executives thinking otherwise
3. Although the industry realizes the importance of IT strategies to achieve innovations, this has not yet been achieved
4. The importance of aligning IT investment to business process management/ re-engineering is highly recognized but is not yet being practiced
5. Industry recognizes the need for IT skills and competences within organizations, but has not yet utilized it for innovation
6. IT investments are driven by "value" but are inhibited by the state of readiness of organizations
7. Industry strongly believes in investigating new technologies for competitive advantage but has not yet taken advantage

Finally, the study clearly demonstrated that there is a big gap between "what the industry think needs to be done" to achieve IT-based innovation and competitive advantage and "how best to achieve it." However, today's 'thinking' of Chief Executives and IT/Innovation Directors has the potential to lay the foundation for the establishment of an industry at a high level of IT maturity.

So, How Can This Issue be Addressed?

The main attributes of the high percentage of systems failure are rarely technical in origin. They are mainly related to the organization – which focuses more on the delivery of technology. They believe that when the technology is in place, it will work its magic. Organizations simply neglect organizational elements, which directly contribute to the success or failure of the new ICT project such as people and process.

The most difficult issues are those dealing with changing people's attitude and business processes, also known as soft issues. The inability to assess organizational readiness to successfully embrace new systems into their work environment will result in wasting time and resources, or even worse, may lead to ICT failure. Lack of attention is another important factor contributing to organizational readiness, and understanding the degree to which an organization may be ready or prepared to obtain benefits from ICT. The Alshawi et al (2008) report revealed that an overwhelming majority of executives and IT directors recognized the importance of e-readiness.

The term "e-readiness" is coined to measure how "ready" organizations are to adopt and use ICT to improve their work practices and performance so that they can achieve sustainable competitive advantage. It reflects the organizational soft issues such as business processes, management structure, change management, people, and culture.

The question is "How could organizations best adopt e-readiness to obtain optimal benefits from existing ICT infrastructure and future ICT investment?" To begin with, organizations need to be process-led and not technology-led. Organizations must not implement new technology directly into current processes but allow technology to be absorbed into "adapted" processes, and to allow an organization to assess itself, if it is ready to accept the technology, or not. Organizational change is needed to create an environment that fosters innovation, clarifies communication, and truly integrates processes.

The above suggests the need to plan organizational change in a proactive manner. In other words, managers should take the initiative to oversee what and how much change is needed prior to ICT implementation. The process of organizational change is always problematic for the organization, particularly when changes are related to the introduction of new ICT (Appelbaum, et al, 1998). Karake (1994) identifies that major changes in IT/IS profoundly affect the organizational aspects (people, processes, structures, and strategies). Katzenbach (1996) and Smith (2005) suggest the hardest things when dealing with organizational change lies in changing the "people system" that includes the organization's structures, planning and control systems, job specialization, training and education programs, degree of centralization, delegation, and participation (Volberda, 1992).

To reduce the resistance to change, by creating a sense of urgency to change and improve communication, it is essential that prior to the introduction of new ICT, the organization needs to determine levels of readiness for change by measuring internal capabilities (Beckard & Harris, 1987; Schein, 1990; Appelbaum, et al., 1998; Smith, 2005). By knowing the level of readiness, an organization can plan ahead to successfully implement the new system (Appelbaum, et al., 1998). Failure to assess organizational readiness prior to the ICT implementation may result in mangers spending more time dealing with the resistance to change, or even worse, may result in ICT failure (Smith, 2005).

The E-Readiness Model

The model developed is of a normative type, and has been created specifically to address a readiness issue, which would increase the likelihood of the success of ICT at organizational level. The model can provide managers with guidelines to prepare the work environment and make it ready to successfully implement new ICT systems. (The full model is presented in Chapter 2).

The e-readiness model is based on the maturity concept. Table 1 lists those models and authors that were referenced and used to create the various elements of

Table 1. The resources used to create the e-readiness model

Concepts/Element/Attributes/Description/Factors	Sources
Model concept • Comprises six levels of maturity • Contains description of each level	• Capability Maturity Model (CMM) (Paulk et al, 1993) • General Practitioner Information System (GPIS) Model (AlSalah, 2002)
Model elements • Comprises four elements; IT Infrastructure, Process, People, Work Environment	• General Practitioner Information System (GPIS) Model (AlSalah, 2002)
Model Attributes • Top Management Perception	• Expert views
Model Attributes • Systems and Communication, Skill, Roles & Responsibility of IT Staff, User Involvement, Organizational Behaviour, IT Policy, Leadership	• General Practitioner Information System (GPIS) Model (AlSalah, 2002) • Nolan (1979) • Earl (1989) • Weil & Broadbent (1998) • A Stages of IT Growth Model (Sutherland & Galliers, 1989)
Model Attributes • Process	• General Practitioner Information System (GPIS) Model (Saleh & Alshawi, 2005) • Capability Maturity Model (CMM) (Paulk et al, 1993)
Model Factors • Drivers • System requirements definition	• IT/IS expertise • Currie & Willcocks (1998) • Tallon et al. (2000) • Cline and Guynes (2001)
Model Factors • Focus • Network Communication	• Tallon et al. (2000) • Cline and Guynes (2001)
Model Factors • Process	• Peppard and Rowland (1995) • Capability Maturity Model (CMM) (Paulk et al, 1993)
Model Factors • Skills	• Nelson (1991)
Model Factors • Capability building	• Crossan et al. (1999)

continued on following page

Table 1. continued

Concepts/Element/Attributes/Description/Factors	Sources
Model Factors • Position of IT/IS Head • Roles	• General Practitioner Information System (GPIS) Model (AlSalah, 2002) • IT/IS expertise
Model Factors • User Involvement	• General Practitioner Information System (GPIS) Model (AlSalah, 2002) • IT/IS expertise
Model Factors • Characteristics	• Leek (1997) • Ibbott & O'Keefe (2004) • Gunasekaran (2004)
Model Factors • IT/IS activities control	• Peterson, (2003) • Karake (1994) • IT expertise
Model Factors • Communication • Participation	• Project Management Maturity Model (Crawford, 2002) • IT/IS expertise

e-readiness, i.e. the readiness models are extracted, adopted, combined, and modified from these models (Saleh, 2007).

Overview of this Book

This book is a compilation of IT/IS readiness models, cases, and associated issues. The rest of the book is divided into two sections; Section 1 presents the e-readiness model and is a compilation of case studies associated with IT/IS readiness models, and Section 2 is a compilation of papers related to different associated issues in IT/IS readiness. Section 1 consists of four case studies. The first case study presents the implementation of the readiness model in higher education sector, the second case study presents an IT/IS readiness assessment for a construction organization, the third case study is about the assessment of e-readiness of a banking organization, and the fourth is an e-readiness assessment of a government service organization.

Section 2 of the book consists of five chapters. The first chapter, by Rehak and Grasseova, is focused on assessing the factors of the external environment in the area of security of information systems in the organization through SWOT analysis. The chapter presents a semi-quantitative assessment of threats in the area of information systems of the organization, evaluates the risks and the assessment of opportunities, and evaluates the overall benefits. The second chapter, by Effah and Light, helps improve our understanding of a small e-support firm's response to the local e-readiness and the global e-business environment in a developing country's context. The third chapter, by Ahmed and Abuelmaatti, presents the findings of a case study that assess the readiness of small and medium enterprises for the implementation of collaborative environments. The fourth chapter of the book presents a benchmarking analysis of the implementation of e-readiness of 100 case studies. The final chapter of the book, by Mohammadian and Jentzsch, presents a fuzzy cognitive mapping approach to provide the capability to capture and represent complex relationships in an IT management process model.

Mustafa Alshawi
University of Salford, UK

Mohammed Arif
University of Salford, UK

REFERENCES

Al-Rashid, W., Alshawi, M., & Al-Mashari, M. (2009). *Exploring Enterprise Resource Planning (ERP) from stake-holder-perspective – Analysis of case study*. In the 4th European Conference in Operations Research, Salford University, Manchester, UK.

Al Saleh, Y. (2002). *IS/IT success and evaluation: A general practitioner model*. PhD Thesis, School of the Built Environment, The University of Salford

Alshawi, M., Khosrowshahi, F., Goulding, J., Lou, E., & Underwood, J. (2008). *Strategic positioning of IT in construction: An industry leader's perspective*. Construct IT For Business, 2008. ISBN 9-781905-732449

Applebaum, S., & St-Pierre, N. (1998). Strategic organisational change: The role of leadership, learning, motivation and productivity. *Management Decision, 36*(5), 289–301. doi:10.1108/00251749810220496

Beckard, R., & Harris, R. (1987). *Organizational transitions: Managing complex change*. Reading, MA: Addison Wesley.

BERR. (2008). *Enterprise: Unlocking the UK's talent. HM Treasury*. UK Government.

Cline, K. C., & Guynes, S. (2001). The impact of Information Technology investment on enterprise performance: a case study. *Information Systems Management, 18*(4), 70. doi:10.1201/1078/43198.18.4.20010901/31467.8

Crawford, J. K. (2002). *Project management maturity model: Providing a proven path to project management excellence*. New York, NY: PM Practices Centre for Business Practices.

Crossan, M., Lane, H. W., & White, R. E. (1999). An organizational learning framework from institution to institution. *Academy of Management Review*, *24*(3), 522–537.

Currie, W. L., & Willcocks, L. P. (1998). Analysing four types of IT sourcing decisions in the context of scale, client/supplier interdependency and risk mitigation. *Information Systems Journal*, *8*, 119–143. doi:10.1046/j.1365-2575.1998.00030.x

Earl, M. J. (1989). *Management strategies for Information Technology*. New York, NY: Prentice-Hall.

Gunasekaran, A. (2004). Editorial: E-commerce enabled manufacturing operations: Issues and analysis. *Information Systems Journal*, *14*, 87–91. doi:10.1111/j.1365-2575.2004.00164.x

Ibbott, C. J., & O'Keefe, R. M. (2004). Trust planning and benefits in a global organizational system. *Information Systems Journal*, *14*, 131–152. doi:10.1111/j.1365-2575.2004.00167.x

Karake, Z. (1994). A study of Information Technology structure – Firm ownership and managerial characteristics. *Information & Management*, *2*(5), 21–30.

Katzenbach, J. R. (1996). Real change management. *The McKinsey Quarterly*, *1*, 148–163.

Leek, C. (1997). Information Systems frameworks and strategy. *Industrial Management & Data Systems*, *97*(3), 86–89. doi:10.1108/02635579710173149

Lientz, B., & Larson, L. (2004). *Manage IT as a business: How to achieve alignment and add value to the company*. Burlington, MA: Elsevier.

Nelson, R. R. (1991). Educational needs as perceived by IS and end-user personnel: A survey of knowledge and skill requirements. *Management Information Systems Quarterly*, *15*, 502–525. doi:10.2307/249454

Nolan, R. L. (1979). Managing the crisis in data processing. *Harvard Business Review*, (March-April): 115–126.

Paulk, M. C., Curtis, B., Chrissis, M. B., Weber, C. V., & Miller, T. R. (1993). *Capability maturity model for software*, version 1.1. (Technical Report CMU/SEI-93-TR-024 ESC-TR-93-177).

Peppard, J., & Rowland, P. (1995). *The essence of business process reengineering*. London, UK: Prentice Hall.

Peterson, R. R. (2003). Information strategies and tactics for Information Technology governance. In Van Grembergen, W. (Ed.), *Strategies for Information Technology governance*. Hershey, PA: Idea Group Publishing. doi:10.4018/978-1-59140-140-7. ch002

Saleh, H. (2007). *Measuring organisational readiness prior to IT/IS investments*. PhD Thesis, School of the Built Environment, The University of Salford.

Saleh, H., & Alshawi, M. (2005). An alternative model for measuring the success of IS projects: The GPIS model. *The Journal of Enterprise Information Management, 18*(1), 47–63. doi:10.1108/17410390510571484

Schein, E. H. (1990). Organizational culture. *The American Psychologist, 45*, 109–119. doi:10.1037/0003-066X.45.2.109

Smith, I. (2005). Achieving readiness for organisational change. *Library Management, 26*(6/7), 27–55. doi:10.1108/01435120510623764

Sutherland, A. R., & Galliers, R. D. (1989). *An evolutionary model to assist in the planning of strategic Information Systems and the management of the Information Systems function*. School of Information Systems Working Paper, Curtin University, Perth, February.

Tallon, P. P., Kramer, K. L., & Gurbaxani, V. (2000). Executives' perceptions of the business value Information Technology: A process oriented approach. *Journal of Management Information Systems, 16*(4), 145–157.

Volberda, H. W. (1992). *Organizational flexibility change and preservation: A flexibility audit and redesign model*. Wolters-Noordhoff.

Weill, P., & Broadbent, M. (1998). *Leveraging the new infrastructure*. Boston, MA: Harvard Business School Press.

Section 1
Cases in IT/IS Readiness

Chapter 1
IT/IS Readiness Maturity Model

Mustafa Alshawi
University of Salford, UK

Hafez Salleh
University of Malaya, Malaysia

EXECUTIVE SUMMARY

This chapter explains the concept of an IT/IS readiness maturity model including particular requirements in terms of four domains, embracing nine attributes: IT infrastructure (top management perception, systems and communication), people (skills, roles and responsibility of IT staff, user involvement), process, and work environment (organization behaviour, IT department, leadership). Each of the attributes consists of 14 factors: top management perception (drivers, systems requirements definition), systems and communication (focus, network communication), skills (type of skills, capability building), roles and responsibility of IT staff (position of IT/IS heads, roles of IT staff), user involvement, process (practices), organizational behaviour (characteristics), IT policy (control of IT/IS activities), and leadership (communication, participation). The following section describes the concept of readiness and maturity, the resources used for element extraction/ adoption and the description of the model.

DOI: 10.4018/978-1-61350-311-9.ch001

DESCRIPTION OF MODEL

The Purpose of IT/IS Readiness Model

A.   The model is intended to be used prior to IS/IT project implementation
B.   The model is a holistic in nature and focuses on soft issues which embrace all the key organizational elements such as IS/IT, people, business processes and work environment.
C.   The model adopts the maturity-level techniques to facilitate the measurement of the "Readiness Gap", i.e. the gap between the current and the required state of readiness, prior to the implementation of a selected IS/IT project.
D.   Each maturity level provides guidelines for managers to improve the readiness status and progress through the maturity levels.

The proposed model is a maturity model composed of six progressive stages of maturity that an organization can achieve in their investment and implementation of IT/IS. These maturity stages are cumulative; which means, in order to get a higher position in the maturity stages, the organization must comply with the pre-ordained requirements for that stage (in addition to those for all the lower stages).

What is being Evaluated?

The model is based on assessing four organizational elements and is shown in Table 1.

Table 1. The characteristics of the key elements of the IT/IS readiness model

Key Elements	Attributes	Characteristics
IT Infrastructure	Top management perception	Describes top management's (business executives) strategic thinking and direction towards the development and utilization of IS/IT in their organizations
	Systems and communication	The development and utilization of applications and the organizations' direction and strategic plan
People	User Involvement	The level of involvement of staff in the IS/IT developments in organizations
	Roles and responsibility of IT staff	The roles and responsibility of IT staff in organizations
	Skills	Acquiring and development of human capacity
Process	Processes	Represented by the process "Practices" within the organization
Work Environment	Organizational behaviour	Actual implementation pattern of IT/IS in organization
	Leadership	The leadership style at both operational and strategic levels
	IT Department	The unit group to provide IT/IS services including infrastructure and applications

A.   IT infrastructure
B.   Business process
C.   People
D.   Work environment

The proposed model is a maturity model composed of six progressive stages of maturity that an organization can achieve in their investment and implementation of IT/IS. These maturity stages are cumulative; which means, in order to get a higher position in the maturity stages, the organization must comply with the pre-ordained requirements for that stage (in addition to those for all the lower stages).

IT Infrastructure

Top Management Perceptions

- **Level 1.** Small IT/IS are developed or purchased where the decision regarding their acquisition tends to be ad-hoc in nature. They are made at low levels within the organization, mainly at group level, and are based on what the management sees taking place within other external organizations.
- **Level 2.** An increased number of IT/IS are being developed or purchased through a small number of IT/IS development plans.
- **Level 3.** Short term development of IT/IS starts to appear with management welcoming user involvement to define needs and requirements.
- **Level 4.** The management begins to consider the long term development of IT/IS with an attempt to align business strategy and IT strategy.
- **Level 5.** The development of IT/IS is used to add value to products or services and to support supply chain activities.
- **Level 6.** The development of IT/IS used to support strategic and innovative business objectives.

System and Communication

- **Level 1.** Most of the systems are small, off-the-shelf financial packages which tend to be independent of each other (i.e. stand-alone) and built/purchased in isolation from other IT/ISs located in the organization or even in the same group.
- **Level 2.** The IT/IS application is more focused on the operational system within the financial area, while a small number of other business-oriented systems are being developed.

3

- **Level 3.** In-house IT/IS applications cover most business operation areas but the IT/IS support varies between the business units. Some new systems are developed, installed, and operated by the central IT/IS department.
- **Level 4.** All required communication protocols such as standard e-mail format is mostly in place, and some EIS, DSS start to appear throughout the organization which indicates the beginning of dependency on the organization-wide network in conducting formal communication.
- **Level 5.** Strategic IT/IS applications are developed with external-oriented data through the use of standard communication such as EDI with external entities such as customers, government and suppliers.
- **Level 6.** The organization uses intra-organizational systems with outside entities (government, suppliers, etc.) with 'sharing' IT/IS services such as the internet, e-commerce technology, etc.

People

Skill

- **Level 1.** The work requires few IT/IS skills. No IT/IS training is provided. User skill is improved by individual effort.
- **Level 2.** IT/IS skills are needed to develop and maintain the system such as programming, analysis and programming skills and being able to install off-the-shelf, ready made packages.
- **Level 3.** Users develop project management skills.
- **Level 4.** Business users start to gain IT understanding. The user gains a proper insight into IT/IS related issues.
- **Level 5.** The business/IT/IS staff gain cross-disciplinary experience. Core technical skills developed. Very knowledgeable IT/IS users.
- **Level 6.** The workgroups are optimizing their IT/IS capability and competency for performing their work processes.

Roles and Responsibility of IT Staff

- **Level 1.** No individual responsible for the IT/IS department. No IT/IS manager. No dedicated IT staff. Small number of low-level technicians and programmers. External contactors may be used to develop/install small systems.
- **Level 2.** The IT/IS manager is responsible for the IT department. Small numbers of IT/IS staff comprising of systems analysts recruited. The IT/IS staff are responsible for adequately understanding the user requirements for systems development.

- **Level 3.** A technically oriented IT/IS manager is appointed. Dedicated IT/IS planners and database administrators are appointed. Adequate technical and specialist staff are to coordinate between current and future IT/IS needs.
- **Level 4.** Apart from programmers, systems analysts and data base administrators, the organization has business analysts. A high level manager for the IT/IS services area is appointed with middle management status.
- **Level 5.** IT/IS managers with senior management status. The organization seeks to develop and retain core hybrid staff.
- **Level 6.** The IT/IS manager becomes a full member of the board of directors and plays an active part in determining strategic direction. The IT/IS staff keep up with the strategic needs of the group.

Users Involvement

- **Level 1.** The IT Department has little control over the users' IT/IS related activities and little user participation in IT/IS decisions.
- **Level 2.** The relationship may exist between users and the IT Department and begin to look at users' knowledge and skills.
- **Level 3.** The IT Department recognizes and welcomes user involvement over the users' needs and requirements concerning IT/IS matters.
- **Level 4.** The users have a significant degree of involvement in IT/IS-related decisions. IT/IS investments are derived from users' stated needs. The IT Department supports the user's activities.
- **Level 5.** Partnerships exist between the IT Department and user groups. The IT Department and users cooperate on an equal basis as partners and continuous striving exists for the integration of organizational workgroups.
- **Level 6.** Central coordination of the strategic coalition between the IT Department and user groups.

Process

- **Level 1.** No standard business process and no alternative plan during crises, and this often leads to compromises on quality. Heavily depends on individual skills to perform the business tasks.
- **Level 2.** Policies and standard procedures established for major business activities.
- **Level 3.** Most business activities are documented and standardized within workgroups.
- **Level 4.** Well-defined business process activities, including standard business descriptions, and models for performing the work tasks within the organization.

- **Level 5.** Established and maintained quantitative objectives for the process about quality and measuring the product/services, the degree of customer satisfaction, and the level of harmony across the supply chain.
- **Level 6.** Ensuring continuous improvement of the process of fulfilling the relevant strategic business goals.

Work Environment

Organizational Behaviour

- **Level 1.** IT/IS is to be used as a tool for performing a single work task
- **Level 2.** IT/IS applications are technology driven
- **Level 3.** IT/IS is one of the many ways to reduce costs in the firm and expenditure on IT/IS is seen as a way to reduce costs
- **Level 4.** IT/IS is vital for streamlining business processes
- **Level 5.** IT/IS is one of the vital elements that lead to competitive advantage
- **Level 6.** IT/IS is the single most critical factor to success in business through knowledge sharing and dissemination

IT Policy

- **Level 1.** Little or no control in IT/IS function and / or no formal IT/IS organizational structure
- **Level 2.** An IT/IS Department is introduced within the organization.
- **Level 3.** The IT/IS Department becomes centralized and IT/IS staff seek control of IT/IS matters.
- **Level 4.** The IT/IS function is well established, and IT/IS services begin to be decentralized with central standards and a policy for co-ordination, implementation, and utility.
- **Level 5.** The central IT/IS department provides an organization-wide communication system, major data processing, and large scale hardware within a large organization. A decentralized management structure exists with the flexibility to support IT/IS initiatives.
- **Level 6.** Coalition between co-ordinated inter-organizational systems

Leadership

- **Level 1.** There is a lack of consistency in the management of IT/IS activities.
- **Level 2.** Management of IT/IS activities with necessary measures and in an ad-hoc manner

- **Level 3.** Management of IT/IS activities with organization wide-policies and standards
- **Level 4.** Management of IT/IS activities with well-defined organizational standards
- **Level 5.** Established management of IT/IS activities with performance measurement
- **Level 6.** Continuous improvement of IT/IS activities management

The Factors/Variables Investigated

A variety of factors need to be achieved in order to get to a higher position in the maturity stages, i.e. the organization must comply with the requirements for that stage (in addition to those for all the lower stages). All factors are contained at each stage to represent their maturity, and the importance of these factors has been discussed in the previous section. Table 2 shows the criteria for each element.

Drivers

These are the factors that drive or trigger the management to make new IT/IS investments. For example, the management decides to invest in the new IT/IS to save operational costs, space reduction, reduce data duplication & redundancy, enhance the company image, etc.

A. **Systems Requirements Definition (SRD):** To identify the requirements of the stakeholders (customers and users) for a new system or proposed system alteration such as *functional requirements* (specific behaviour; what the system

Table 2. The criteria/variables for the IT/IS readiness model

Element	IT Infrastructure		People		
Attributes	Top Management Perception	Systems and Communication	Skill	Roles & Responsibility of IT staff	User Involvement
Factors	• Drivers • Systems requirements definition	• Focus • Network communication	• Type of skills • Capability building	• Positions of IT/IS heads • Roles of IT staff	• User involvement
Element	**Process**	**Work Environment**			
Attributes	Process	Organization Behaviour	IT Policy	Leadership	
Factors	• Practices	• Characteristics	• Control of • IT/IS activities	• Communication • Participation	

can do such as data processing, data manipulation, graphical user interface, ease of use, security etc.) and *non-functional requirements* (to judge the operation of a system such as reliability, performance and cost). SRD is an important part of the system design process, whereby business analysts along with the stakeholders and system developers identify the needs of a client. Among popular techniques used to identify SRD are the stakeholder's interview, workshops, prototypes etc.

B. **Focus:** The utilization of IT for the organization's improvement such as to improve internal efficiency; enhance overall organizational effectiveness; extend geography and market reach; change the industry or market place (Tallon et al, 2000).

C. **Network Communication:** A computer *network coverage* within the organization that is capable of exchanging information electronically between individuals, groups, and business units.

D. **Process:** A workflow, or group of activities (sometimes called methods and procedures), that produce an outcome valued by an internal or external customer'. (Hammer and Champy, 1993; Harrington, 1991). They (a work flow or group of activities) are generally cross-functional and horizontal in nature with no single person having responsibility for the entire process (Innovative Manufacturing Initiative, 1994).

E. **Types of Skills:** A specific range of IT/IS skills needed by the user to maximize the potential of IT systems and meet the requirements of a specific job.

F. **Capability Building:** The mechanism for an organization improving their learning capabilities of IT/IS skills (Murray and Donegan, 2003).

G. **Position of IT/IS Head:** The position of the IT/IS Head in the organization structure.

H. **Roles of IT Staff:** The roles and responsibility of IT staff including technicians, programmers, analysts, managers, etc.

I. **User Involvement:** The approach of user involvement in the IT/IS development process by an individual or members of the target user group.

J. **Organization Behaviour Characteristics:** The actual implementation pattern of IT/IS within working environments in the organization.

K. **Control of IT/IS Activities:** The pattern of enterprise decision making authority on IT resources (infrastructure and application) among IT departments and business units.

L. **Communication:** The communication planning throughout the organization at all levels of the management regarding the IT/IS implementation/activities.

M. **Participation:** The participation of top and middle management in the IT/IS initiatives.

The summary of IT/IS readiness model can be found in Tables 3, 4, 5, and 6.

Table 3. Summary of research model – Information Technology infrastructure element

	Top Management Perceptions *Top management (business executive's) strategic thinking and direction towards the development and utilization of IT/IS in their organizations*	**Systems and Communication** *The development and utilization of applications and systems to facilitate the organization's direction and strategic plan*
Level 6	**STRATEGIC OBJECTIVES** *The development of IT/IS used to support strategic and innovative business objectives.* • **Drivers:** Global competition. • **System Requirements Definition:** Full in-house with minimum intervention from company's partners.	**INTRA-ORGANIZATIONAL SYSTEMS** *Organization uses intra-organizational systems with outside entities (government, suppliers, etc) sharing IT/IS services such as the internet, e-commerce technology etc.* • **Focus:** Managing information for strategic business core-capabilities. • **Network Communication:** Intra-Organizational networking.
Level 5	**ADDED VALUE** *The development of IT/IS is used to add value to products or services and to support supply chain activities.* • **Drivers:** Partner's supply chain. • **System Requirements Definition:** Mainly in-house with intervention from supply chain partners and vendors.	**STRATEGIC SYSTEM** *Strategic IT/IS applications are developed with external-oriented data through the use of standard communication such as ERP, SCM, EDI with external entities such as customers, government and suppliers.* • **Focus:** Managing supply-chain. • **Network Communication:** Supply-chain with partners.
Level 4	**LONG TERM BUSINESS INVESTMENT** *The management start to consider long term development of IT/IS with an attempt to align business strategy and IT strategy.* • **Drivers:** Business process improvement. • **System Requirements Definition:** Mainly in-house with intervention from vendor.	**COMMUNICATION PROTOCOLS** *All required communication protocols such as standard e-mail format is mostly in place and some EIS, DSS start to appear throughout the organization which indicates the beginning of dependency on the organization-wide network in conducting formal communication.* • **Focus:** Manipulating information to assist managers in decision making. • **Network Communication:** Organizational-network systems.
Level 3	**SHORT TERM BUSINESS INVESTMENT** *Short term development of IT/IS start to appear with management welcoming user involvement to define needs and requirements.* • **Drivers:** Organization communication. • **System Requirements Definition:** Mainly in-house with intervention from vendors.	**WORK GROUP SYSTEM** *In-house IT/IS applications cover most business operational areas but the IT/IS support varies between the business units. Some new systems are developed, installed, and operated by the central IT/IS department.* • **Focus:** Full integration of application organizational-wide. • **Network Communication:** Organizational-network systems.
Level 2	**PROJECT-ORIENTED INVESTMENT** *An increasing number of IT/IS being developed or purchased through small number of IT/IS development plans.* • **Drivers:** Work task requirements. • **System Requirements Definition:** Partially from in-house and mostly form vendor.	**ADMINISTRATION SYSTEM** *The IT/IS application focus is more on the operational system within the financial area while a small number of other business-oriented systems are being developed.* • **Focus:** Co-ordination information activities. • **Network Communication:** Networking within business unit.

continued on following page

Table 3. Continued

	Top Management Perceptions *Top management (business executive's) strategic thinking and direction towards the development and utilization of IT/IS in their organizations*	Systems and Communication *The development and utilization of applications and systems to facilitate the organization's direction and strategic plan*
Level 1	**AD-HOC INVESTMENT** *Small IT/ISs are developed or purchased where the decision regarding their acquisition tend to be ad-hoc in nature, made at low levels within organization, mainly at group level and are based on what the management sees taking place within other external organizations.* • **Drivers:** Copying/duplication. • **System Technical Requirements Definition:** Heavily depends on vendor.	**STAND-ALONE SYSTEM** *Almost all systems are small and off-the-shelf financial packages, which tend to be independent of each other (stand-alone) and built/purchased in isolation from other IT/ISs located in the organization or even in the same group.* • **Focus:** Financial activities. • **Network Communication:** Stand-alone.

Table 4. Summary of research model – Process element

	Process *Represented by the process "practices" within the organization*
Level 6	**INSTITUTIONALISE AN OPTIMISING PROCESS** *Ensuring continuous improvement of the process in fulfilling the relevant strategic business goals.* • The entire supply chain is focused on continuous business process improvement (strength, weakness and evaluation).
Level 5	**INSTITUTIONALISE QUANTITATIVELY MANAGED PROCESS** *Establish and maintain quantitative objectives for the process of quality and measuring the product/services, the degree of customer satisfaction, and the level of harmony across the supply chain.* • The capability to set quality goals and measures on business process activities.
Level 4	**INSTITUTIONALISE DEFINED PROCESS** *Well-defined business process activities including standard business descriptions and models for performing the work tasks within the organization.* • A well-defined business process includes standard descriptions and models for performing work.
Level 3	**INSTITUTIONALISE MANAGED PROCESS** *Most business activities are documented and standardized within workgroups.* • Business process is documented, standardized and integrated organization-wide.
Level 2	**ACHIEVE SPECIFIC GOALS** *Policies and standard procedures established for major business activities.* • Business process scope is identified and improved as the work progresses.
Level 1	**AD-HOC LEVEL** *No standard business process and no alternative plan during crises which often leads to compromises on quality. Heavily dependant on individual skills to perform the business tasks.* • Business process is unpredictable and constantly changed or modified as the work progresses.

Table 5. Summary of research model – People element

	Skill *The acquisition and development of human capacity*	**Roles and Responsibility of IT Staff** *The roles and responsibility of IT staff in organizations.*	**Users' Involvement** *The level of involvement of staff in the IS/IT developments in organizations*
Level 6	**SKILLS ENHANCEMENT** *The workgroups are optimizing their IT/IS capability and competency for performing their work processes.* • **Type of skills:** Acquiring skills to develop IT core capabilities. • **Capability Building:** Intra-organizational sharing experience.	**EXECUTIVE BOARD STAFF** *The IT/IS manager becomes a full member of the board of directors and plays an active part in determining strategic direction. The IT/IS staff keep up with the strategic needs of the group.* • **Position of IT/IS Head:** IT Manager with full membership of the board of directors. • **Roles:** To align IT/IS strategy and business strategy.	**STRATEGIC** *Central coordination of the strategic coalition between the IT Department and user groups.* • **User Involvement:** User group establishes their role and become central reference in IT/IS project team looking at aligning IT/IS strategy and business strategy.
Level 5	**SKILLS INTEGRATION** *The business/IT/IS staff gains cross-disciplinary experience. Core technical skills developed. Very knowledgeable IT/IS users.* • **Type of skills:** Gains cross-disciplinary experience to support supply-chain activities. • **Capability Building:** Knowledge sharing and knowledge management (multi-disciplinary).*	**HYBRID STAFF** *IT/IS manager with senior management status. The organization seeks to develop and retain core hybrid staff.* • **Position of IT/IS Head:** IT Manager with senior management status. • **Roles:** To set-up IT/IS strategy	**PARTNERSHIP** *Partnership exists between IT Department and user groups. IT Department and users cooperate on an equal basis as partners and continuously strive to integrate the organizational workgroups.* • **User Involvement:** User group become permanent in the IT/IS project team looking at IT/IS strategy.
Level 4	**SKILLS DEPLOYMENT** *Business users start to gain IT understanding. A user gains a proper insight into IT/IS related issues.* • **Type of skills:** Using IT/IS to make decisions. • **Capability Building:** Centrally integrated training program within the organization.	**BUSINESS-ORIENTED STAFF** *Apart from programmers, systems analysts and data base administrators, the organization has business analysts. A high level manager for the IT/IS services area is appointed with middle management status.* • **Position of IT/IS Head:** IT Manager with middle management status. • **Roles:** To manage information across organization.	**CONSULTATIVE** *The users have a significant degree of involvement in IT/IS-related decisions. IT/IS investments are derived from users' stated needs. The IT Department supports user's activities.* • **User Involvement:** Focus group consultation on managing information across organization.

continued on following page

Table 5. Continued

	Skill *The acquisition and development of human capacity*	**Roles and Responsibility of IT Staff** *The roles and responsibility of IT staff in organizations.*	**Users' Involvement** *The level of involvement of staff in the IS/IT developments in organizations*
Level 3	**IT/IS PROJECT MANAGEMENT SKILLS** *Users developed project management skills.* • **Type of Skills:** Considerable technical and project management skills. • **Capability Building:** Centrally integrated training programs within organizations.	**TECHNICAL-ORIENTED STAFF** *A technically oriented IT/IS manager is appointed. Dedicated IT/IS planners and database administrators are appointed. Adequate technical and specialist staff to coordinate between current and future IT/IS needs.* • **Position of IT/IS Head:** Technical IT Manager. • **Roles:** To set-up purchasing policy and centralized IT/IS activities.	**REPRESENTATIVE** *The IT Department recognizes and welcomes user involvement on the users needs and requirements concerning IT/IS matters.* • **User involvement:** Focus group consultation on purchasing and centralized IT/IS activities.
Level 2	**IT/IS SKILLS DEVELOPMENT** *IT/IS skills are needed to develop and maintain systems such as programming, analysis and programming skills and being able to install off-the-shelf, ready made packages.* • **Type of skills:** Purely technical skills. • **Capability building:** Team based/effort within workgroups.	**LIMITED ROLE** *The IT/IS manager is responsible for the IT department. Small numbers of IT/IS staff comprised of systems analysts recruited. The IT/IS staff are responsible for adequately understanding the user requirements for systems development.* • **Position of IT/IS Head:** IT Manager at IT department. • **Roles:** Providing technical support.	**AD-HOC INVOLVEMENT** *A relationship may exist between users and the IT Department and begin to look at users' knowledge and skills.* • **User Involvement:** Individual consultations on technical support.
Level 1	**BASIC IT/IS SKILLS** *The work requires few IT/IS skills. No IT/IS training provided. User skill is improved by individual effort.* • **Type of skills:** Basic IT/IS skills. • **Capability Building:** Individual based/effort.	**NO ROLE** *No individuals responsible for the IT/IS department. No IT/IS manager. No dedicated IT staff. Small numbers of low-level technicians and programmers. External contactors may be used to develop/install small systems.* • **Position of IT/IS Head:** No IT Manager. • **Roles:** No role.	**BASIC INVOLVEMENT** *IT Department has little control over users IT/IS related activities and little user participation in IT/IS decision.* • **User Involvement:** No user participation in IT/IS improvement/enhancement.

Table 6. Summary of research model – Work environment element

	Organization Behaviour *Actual implementation pattern of IT/IS in an organization*	**IT Governance** *The unit group to provide and control IT/IS services including infrastructure and application*	**Leadership** *The leadership communication and participation level*
Level 6	**KNOWLEDGE - CULTURE** *IT/IS is the most single critical factor to success in business through knowledge sharing and dissemination.* **Characteristics:** • IT benefit is measured quantitatively (financial cost/profit ratio) and qualitatively (usefulness for all members of the company). • Information is shared and exchanged for product and process innovation and improvements, decision making and organizational adaptation and renewal. • 'Informational culture' - design the system according the explicit and implicit knowledge that individuals possess.	**ESTABLISHED HYBRID MODEL/FEDERAL STRUCTURE** *Coalition between co-ordinated inter-organizational system.* • **IT/IS Activities Control:** Centralized IT/IS infrastructure and decentralized application to support intra-organizational IT/IS activities.	*Continuous improvement of IT/IS activities management.* • **Communication:** Continuous improvement of the communication process. • **Participation:** Participates actively in the continuous improvement of IS/IT projects such as using the data obtained from efficiency and effectiveness matrices for strategic decision making.
Level 5	**CAPABILITY APPROACH** *IT/IS is one of the vital elements that lead to competitive advantage.* **Characteristics:** • IT/IS used to provide better products/services to customers. • Staff encouraged to input IT/IS driven ideas to improve products/services. • IT/IS investment influenced by competition.	**HYBRID MODEL/FEDERAL STRUCTURE** *Central IT/IS departments provide an organization-wide communication system, major data processing, and large scale hardware within large organizations. A decentralized management structure exists with the flexibility to support IT/IS initiatives.* • **IT/IS Activities Control:** Centralized IT/IS infrastructure and decentralized application to support supply chain activities.	*Established management of IT/IS activities with performance measurement.* • **Communication:** Well documented and integrated communication planning for overall organizational structure. • **Participation:** Participates in performance measurement of IS/IT projects such as efficiency and effectiveness.
Level 4	**ORGANIZATION - APPROACH** *IT/IS is vital for streamlining business processes.* **Characteristics:** • The focus is on ensuring all organization functions and operations are running smoothly. • Process oriented. • Sharing processes for particular tasks within the organization.	**ESTABLISHED IT DEPARTMENT** *IT/IS function is well established and IT/IS services begin to decentralize with central standards and a policy for co-ordination, implementation and utility.* • **IT/IS Activities Control:** Centralized IT/IS infrastructure and decentralized application (policy).	*Management of IT/IS activities with well-defined organizational standards.* • **Communication:** A communication plan is expected for all activities. • **Participation:** Participation of all IS/IT activities within the organization such as project prioritization, change control, risk identification etc.

continued on following page

Table 6. Continued

	Organization Behaviour *Actual implementation pattern of IT/IS in an organization*	**IT Governance** *The unit group to provide and control IT/IS services including infrastructure and application*	**Leadership** *The leadership communication and participation level*
Level 3	**COST - APPROACH** *IT IS is one of the many ways to reduce costs in the firm and expenditure on IT IS is seen as a way to reduce costs.* **Characteristics:** • IT benefits are measured quantitatively (financial cost/profit ratio). • Profit driven - every project is based on its feasibility to gain profit. • Tight budget control - centralized (top management) monitoring on expenditure.	**IT DEPARTMENT CENTRALISED** *IT IS Department become centralized and IT IS staff seek control of IT IS matters.* • **IT/IS Activities Control:** Centralized purchasing policy and support (policy).	*Management of IT IS activities with organization wide-policies and standards.* • **Communication:** There are an organization-wide policy and standards for communications. • **Participation:** Participation on large scale and high cost IS/IT activities such as cost and schedule etc.
Level 2	**TECHNOLOGY - APPROACH** *IT IS applications are technology driven.* **Characteristics:** • Technology-led. • 'Informatics culture' - designs the system from technical point of view, then persuades staff to adapt to it. • IT success is measured in IT terms rather than the impact made on the business.	**IT DEPARTMENT RECOGNISED** *IT IS Department is introduced within the organization.* • **IT/IS Activities Control:** Technical support.	*Management of IT IS activities with necessary measures and in an ad-hoc manner.* • **Communication:** No established standards in place for communications planning. • **Participation:** Ad-hoc participation basis by individual managers such as IS/IT project status (milestone - plan and actual) etc.
Level 1	**AD-HOC APPROACH** *IT IS is thought of as a tool for performing single work tasks.* **Characteristics:** • Users determine their needs and requirements. • No data sharing culture exists. • Each individual has their own separate processes.	**NO FORMAL IT DEPARTMENT** *Little or no control in IT IS functions and or no formal IT IS organizational structure.* • **IT/IS Activities Control:** No control on policy.	*Lack of consistency in the management of IT IS activities.* • **Communication:** No communication. • **Participation:** No participation.

REFERENCES

Hammer, M., & Champy, J. (1993). *Reengineering the corporation: A manifesto for the business revolution*. New York, NY: Harper Business.

Harrington, H. J. (1991). *Business process improvement*. US: McGraw-Hill.

Murray, P., & Donegan, K. (2003). Empirical linkages between firm competencies and organisational learning. *The Learning Organization*, *10*(1), 51. doi:10.1108/09696470310457496

Tallon, P. P., Kraemer, K. L., & Gurbaxani, V. (2000). Executives' perceptions of the business value of information technology - A process-oriented approach. *Journal of Management Information Systems*, *16*(4), 145–173.

Chapter 2

Readiness in Systems Implementation:
Lessons from the Higher Education Sector

Eric Lou
University of Salford, UK

Hafez Salleh
University of Malaya, Malaysia

EXECUTIVE SUMMARY

Higher education sector is notorious for lagging behind the industrial sector in the application of IT/IS systems and infrastructure. This chapter presents the application of the IT/IS readiness model in a higher education organization. This organisation was established in 1967 and currently has about 2,500 staff and 18,000 students, of which, 3,000 are international students from all over the world. The organization comprises of 14 schools and 13 research institutes and offers programmes various fields, which include virtual reality, magnetic and optics, business, law, genetic algorithms, health-related studies, and building construction. In 1996, Academic Division (AD) identified the need to improve the management of the student database due to the increase of students and programs offered by the organization. AD also identified that the Legacy Student Information System (SIS) was unable to cope with the increasing demand of data administration.

DOI: 10.4018/978-1-61350-311-9.ch002

This case study presents the overview of issues encountered while assessing the e-readiness of the organisation after most of the systems went live. Post implementation, the system has been able to reduce the redundancies in processes and has been able to provide a more effective support to students and staff. However, still there are several issues and conflicts that need to be resolved, and a radical rethink of the processes supporting the IT system is needed to achieve any further efficiency.

BACKGROUND

This chapter presents the application of the IT/IS readiness model in a higher education organization, which we will refer to as Organization A. The organization was established in 1967 which Prince Philip, the Duke of Edinburgh, as the first Chancellor. To date, the organization has about 2,500 staff and 18,000 students, of which 3,000 are international students from all over the world. The organization comprises of 14 schools and 13 research institutes and offers programmes various fields, which include virtual reality, magnetic and optics, business, law, genetic algorithms, health-related studies and building construction.

The Governing Court (Board of Trustees) of this organization has some 200 members amongst whom are representatives of educational, professional, regional and national bodies. The responsibility of the Court includes the appointment of the Chancellor and the Vice-Chancellors, and certain members of the Council. The Council is the business executive body of Organization A and is responsible for its overall organizational structure and for the management and administration of the finances and property of the institution. The Council is composed of some 37 academic, non-academic, student, and lay members. The Senate is composed wholly of academic members and is the academic authority of Organization A. Its duties are to promote the academic work of the organization, both in teaching and research, and to regulate and superintend the education and discipline of students. The Vice-Chancellor, Pro-Vice-Chancellors, Deans of Faculty, Associate Deans, Heads of School, and the Directors of Academic Enterprise, Graduate Studies and Information Services are all ex-officio members of the Senate.

Sequence of Events

In 1996, Academic Division (AD) identified the need to improve the management of the student database due to the increase of students and programs offered by the organization. AD also identified that the Legacy Student Information System (SIS) was unable to cope with the increasing demand of data administration. Each of Faculties/Schools/Programmes managed their own data according to their requirements.

The exchange of student information between Faculties/Schools/Programmes and AD and other support services was mainly through hard copy. The Legacy SIS used was developed in an uncoordinated manner, reflecting interests in different areas, and resulting in issues of redundancy and inefficiency. For example, AD, Library, Faculty and Schools may have to separately update the same information about a student. The Legacy SIS was developed mostly to meet the needs and requirements of AD; and there were high operating and administration costs for delivering the same administration function across the organization due to duplication and redundancy work practice. Over the next 2 years, the decision to switch to the new system was made by the Senior Management Team to replace the internally developed Legacy SIS. Information System Division (ISD), the owner of all information services on campus was assigned to do a feasibility study on the potential systems available, and decided to go to the third party package called System X. This is a product from a vendor based in the United States of America.

System X was purchased in August 1999; which is a centralized Database Management System (DBMS) running on YYY to store all student information and their academic career and used by all departments across the organization (Figure 1).

The system is Unix-based, and uses the YYY relational database management system. System X is a tool to manage the administration of student affairs across a life cycle of study and begins with the enquiries, the students' application for enrolling in Organization A courses, all the way up to the alumni relationship after graduating. System X deals with all types of student stages and status, including undergraduate, postgraduate and full time, part-time and distance learning. There are different modules (packages) which are part of system X as shown in Figure 2. However, the organization A only purchased the 'Student' module since the other areas were covered by different existing systems such as 'SAP' for human resource, 'Agresso' for finance, 'Talis' for library management, 'Athens' for resources etc.

Figure 1. System X

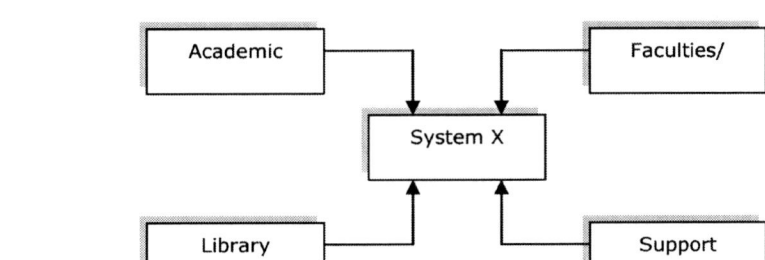

Figure 2. The System X modules

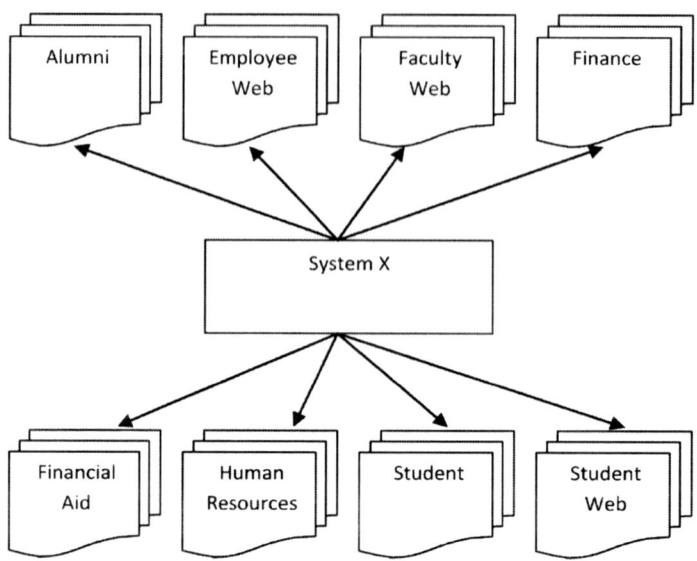

System X went 'live' in September 2000 with undergraduate functions running first. Since then, gradual functionalities have been added to the system. Initially, the organization anticipated that the implementation of System X would go on without any major revision of their business processes. However, when the implementation was taking place, the organization started to realize the needs to make major efforts to customize the system to meet the requirements of the organization.

When the system was originally implemented, ownership belonged to Academic Division (AD), which took care of system delivery and maintenance. However, in 2002, due to unsolved technical problems, the ownership of the system was passed on to the ISD. It is suspected that this system failure symptom occurred because of the focus on technology delivery, such as system functioning. Further investigations have since determined that system X is designed to deal with ad-hoc enquiries, and cannot manage the enquiries and analyze the enquiry data in the way the organization requires. The over-riding problem is that there is no concept of a returning enquirer, and no record is kept on an interest in particular programs or other information offered. It is suspected that System X will not fulfill requirements for the enquiry part of the student life-cycle. This shortfall is due to Organization A's requirements being far more wide-ranging than System X can provide. After this, the organization started to initiate projects to overcome this by introducing Stabilization Project in 2002 to identify key systems enhancements, and to support systematic business analysis of the organization's administrative processes and to implement system changes once identified.

Figure 3. System X managing student life-cycle

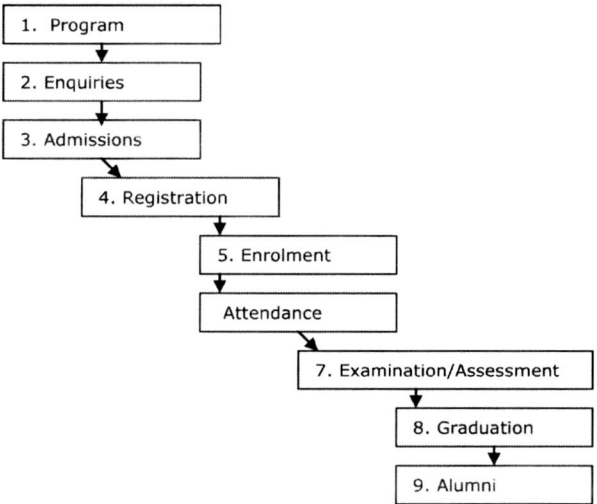

In August 2003, as part of the stabilization project, a workshop was organized to identify all the stages of the student lifecycle, and its processes that needed to be mapped to System X's functionality. A number of staff from different departments were invited to take part in this workshop. The workshop identified nine processes involved in the student lifecycle, i.e. program definition, enquiries, admissions, registration, enrolment, attendance/teaching, examination/assessment, graduation and alumni (Figure 3). The workshop identified the issues that needed to be dealt with as follows:

1. System X does not record interest in particular programs, and treats each connection to the system as independent, so no history is recorded. There is no such concept as a returning enquirer - the organization needs to target the available resources on enquirers who have a high prospect in applying for the organization's programs. The organization needs to identify these high prospects by recording the relevant information, such as interest in programs, frequency of looking at the organization's information, where they currently live, and the previous institution attended.
2. System X does not hold the concept of category - when a user is searching for a program they will want to search by category, i.e. engineering courses or music related, etc.
3. The additional information cannot be customized to the specific needs of the enquirer. When an enquirer views additional information they should only be presented with the relevant information, which will depend on factors such as the level of study and the country of origin.

4. System X does not hold full details on programs – when a program is identified by an enquirer, they seek more details about it. This includes a detailed description about what the program offers, and advertised entry requirements. This information is listed in the printed prospectus, but needs to be online.

5. No history is kept on enquirers. As each enquiry is treated independently from all others there is no such concept as a returning enquirer, which in turn means that no history is kept, and hence no management of that enquirer is possible.

6. System X does not distinguish between a corporate user and an individual enquirer. System X needs to treat corporate entities differently, so that they can be sent items automatically, i.e. a prospectus, regular information on open days, etc.

7. System X does not hold all the required information on enquirers. In order to run the business processes the organization requires additional information to be stored over and above the information held by system X.

•   Also under the stabilization project, a dedicated committee for System X development was formed as shown in Figure 4

The SIS Project Executive Group (SISPEG) has overall responsibility for the Project, and all other committees associated with the Project report to it. The Executive Group meets every two months. Members of SISPEG comprise of the Pro-Vice-Chancellor Institutional and Student Services (Chair); Director of External Relations; Associate Director Information Systems, ISD; Director of Information Services; Associate Dean (Teaching and Learning); Chief Accountant, Finance Division; Project Delivery Manager, Information Systems, ISD; Academic Registrar; Chair of the User Council; User Council Representative; School and Faculty Representative; Registrar and Secretary; and the SIS Service Delivery Manager. The SIS User Council (SISUC) acts as a forum for SIS Senior Contacts and representatives of central administrative units to review how System X is currently used to

Figure 4. System X's project committee

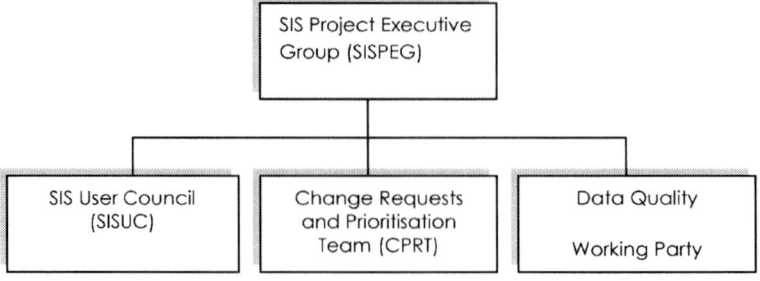

support their work, to review proposed SIS developments, and to recommend future developments. The membership of SISPUC comprises of faculty representatives, school representatives, support services representatives and the System X project team.

The Change Requests and Prioritization Team (CPRT) meets every 3 months and considers all operational matters relating to the Project, making recommendations, where appropriate, to the Project Executive Group. Members of CRPT comprise of:

1. Service Delivery Manager, SIS Project (Chair)
2. SIS User Council Representative
3. External Relations
4. Finance Division
5. Undergraduate Office, Academic Division
6. Project Delivery Manager, Information Systems
7. Business Analyst, SIS Project

The Data Quality Working Party [DQWP] reports to the Student Information Systems Project Executive Group [SISPEG]. The DQWP should recommend data quality policy and, through sub-groups where appropriate, mechanisms to ensure that data quality is assured. Members of DQEP comprise of:

1. SISPEG Representative (Chair)
2. SIS User Council Representative
3. Academic Division Representative
4. External Relation Representative
5. ISD Representatives
6. Planning and Executive Support Unit Representative

This system was implemented for undergraduates in the year 2007 and currently it is being analyzed for postgraduate students. The SIS Postgraduate (SISPG) Project is focused on the exploitation of existing functionality so as to quickly improve the level of support provided for administering postgraduate students using the central SIS. A dedicated committee for SISPG development was formed with the tasks of initiating the process for the recruitment of an external consultant to lead and manage the project team; identifying major risks and agree action plans to manage these; monitoring project plans and timescales produced by the SISPGPT against agreed project priorities; acting as the escalation point for issues which cannot be resolved by the SISPGPT; and ensuring that the project delivers value for money. Members of SISPGPB comprise of the Pro-Vice Chancellor Research (Chair); Director of Graduate Studies; Academic Registrar; Director of ISD; Research Institute Representative; and Secretary. The SISPGPT is responsible for defining the appropriate business processes, controls, and responsibilities for prioritized work elements, that report to the SISPGPB including work through the prioritized project elements defining appropriate business processes, controls and responsibilities; ensuring that business processes, controls and responsibilities are implemented operationally in a controlled and coordinated manner (including ensuring that communications, training and awareness raising events take place); providing progress reports to the SISPGPB; and raising issues to the SISPGPB as required. Figure 5. Illustrates System X Development Timeline

ANALYSIS AND DISCUSSION

The following section addresses each attribute of the four domains of the readiness model within the context of the organization.

IT INFRASTRUCTURE

Top Management Perceptions

The process of the realization of top management strategic thinking and planning are as follows:

*   The strategic thinking and direction of the organization are generally set by the Executive Group Team, which comprises of the Vice-Chancellor, Pro-Vice-Chancellors, Director of Finance, Registrar and Secretary.

Figure 5. System X development timeline

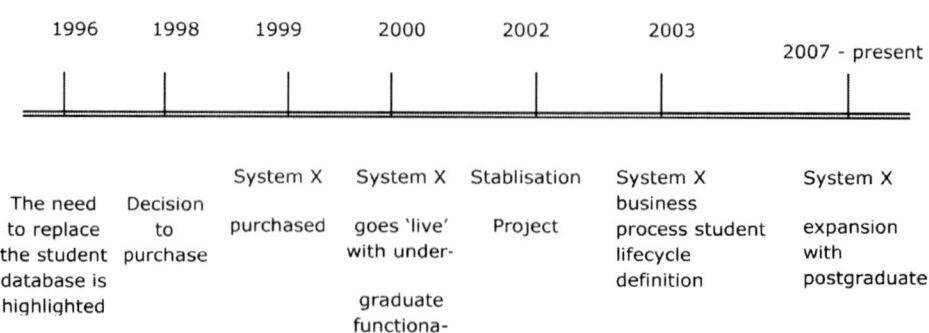

- Then it is presented for the consideration of the Senate and the Council.
- Normally, the Executive Group Team will get a wide consultation of all members of staff including the Senior Management Team which comprise of the Deans of Faculty, the Director of Academic Enterprise and the Director of Information Services
- After the consideration of the Senate and the Council, the Senior Management will plan how to make the strategic thinking and planning become reality.

In the year 2004, The senior management with a wide consultation of all members of staff, launched *'The Strategic Framework 2005-2015'* containing the organization's long term-mission, goals and also priorities for more immediate action. This framework represents the fundamental aims of the organization, which blended the senior management strategic thinking, direction and the insight, understanding and feedback of all members of staff including academic and non-academic staff. The Strategic Framework 2005-2015 seems to be applied as a general guideline for future IT/IS development. The *status prior to project* refers to the period prior to year 2000 where legacy SIS was still in operation. The *current status of the project* refers to when System X came 'live' with undergraduate functionality. The *target status* refers to the readiness level that organizations desire to be at in the future.

Status Prior to Project

Drivers

The development of Legacy SIS is to cater for the basic needs of a student information database in performing daily work tasks in Faculties/Schools/Programmes, and Academic Division and other support services.

Systems Requirements Definition

Organization A used the internal staff in Academic Division to define non functional requirements such as (performance, cost, and reliability) and Academic Information Services (AIS) to define the needs and functional requirements of the Legacy SIS system (data manipulation, security etc). The vendor, with the help of IT staff, developed further details of the functional and non-functional requirements.

Current Status of the Project

Drivers

The decision to develop System X was due to several limitations of legacy SIS - high operating and administration costs for delivering the same administration function across the organization; legacy SIS was designed mostly to meet the needs of AD; the AD, library, faculty and schools have their own student databases and may have to update the same information about a student; and system X increases the level of communication among business units within Organization A, i.e. Academic Divisions, Schools and Faculties, Library Services, Information Service Division, etc., by sharing the same student information for decision making, e.g. student registration and fees.

Systems Requirements Definition

ISD has formed a dedicated team headed by business analysts for the modification, and to determine the functional and non-functional requirements of the organization. An example of functional requirements identified in System X's business process student lifecycle definition workshop is:

*   Allow direct entry of information by remote enquirer via the Internet.
*   Allow data to be entered by administrative staff on behalf of the enquirer either by the Internet or local network.
*   Allow updates to the stored data by the enquirer or authorized administrative staff
*   Allow enquiries to be identified with specific marketing campaigns
*   Identify the source of all enquiries
*   Provide the enquirer information on all levels of teaching & research
*   Provide details on additional supporting information such as accommodation, open days, etc.
*   Allow read-only access to standard reporting tools
*   Record the category for courses the enquirer shows an interest in
*   Record the courses the enquirer is interested in
*   By adding weighted scores against each enquirer, derived from attributes such as country of origin, usage of the system and programmes/modules viewed, identify overseas prospects
*   By adding weighted scores against each enquirer, derived from attributes such as the previous educational establishment and checking this against a list of feeder schools and also the geographical location, usage of the system and programmes/modules viewed, identify home based prospects.
*   Record all brochures sent to an individual.
*   Identify all letters posted to an individual.
*   Record all email types sent to an individual.
*   Manage & Record all responses.
*   Manage enquirers who become applicants.

- When an email address is known & valid, deliver all the correspondence via email.
- For Internet active enquirers, send reminder emails where and when appropriate.
- Manage enquirers who become applicants.
- Remove 'Gone Always' from the system.
- Archive data that is no longer current.
- Remove/merge duplicate records by selective administrative staff.
- Identify organizations such as schools, colleges and agents who make requests on behalf of groups of potential students.
- Send out full information packs to these organizations.
- Manually add organizations to the system and remove organizations that are no longer active.

Examples of non-functional requirements identified in System X business process student lifecycle definition workshop are:

- *Organization*
  - Allow administrative staff to add new letter types to the system and to identify them as single or multiple 'sends'.
  - Allow administrative staff to add new email types to the system and to identify them as single or multiple 'sends'.
  - Ensure the data is available for centralized functions.
  - Ensure the data is available to all sections of Organization A that require it. This will include the schools and faculties, through a single integrated solution.
- *Security*
  - Ability to make an enquiry on-line. Logging onto the system will uniquely identify the enquirer. All correspondence will be sent to the enquirer's email account.
  - All accounts to be password protected.
  - Forgotten passwords can be requested and subsequently sent to the email account of the enquirer.
  - The enquirer will have the option to change their password.
  - Encrypt all passwords stored.
- *Operation*
  - Automate the backing up of the system
  - Automatically inform support staff when the system is unavailable.

- *Resources*
  - ○ Generic emails to be used for all email correspondence. These are to be monitored by the administrative team.
- *Data quality*
  - ○ Ensure the quality of the data is of a standard that can be of use to Organization A.

Target Status (the Readiness Level that Organizations Desire to be at in the Future)

Drivers

The Executive Group and Senior Management Group of Organization A have derived a number of key influencing factors that will drive Organization A's future IT/IS requirements which are published in *"Vision Statement – Vision for ICT Services, 2006"* as follows:

- **Global Competition:** Widening participation and expansion into global markets will drive a need for increased availability and location independence of ICT services, along with the requirement to provide services adapted for local needs.
- **Competitive Advantage:** The introduction of student fees is expected to increase competition within the higher education sector, with students demanding improved quality of service, improved breadth of service, and demonstrable value for money. Increased competition will create an environment in which innovation in service provision can give institutions a significant competitive advantage.
- **Business Process Improvement:** The impact of existing and new regulatory frameworks will increase compliance and auditing requirements for ICT services and will demand improved control over information lifecycles and business processes.

System Requirements Definition

Organization A used its in-house capabilities to develop system requirements during the development of System X. The in-house expertise has the experience to develop its own functional and non-functional requirements. It is now policy that future development of IT systems will be done using internal expertise, with opinions from its customers and vendor.

The results of Top Management Perception analysis are shown in Table 1.

Table 1. IT Infrastructure: Top Management Perceptions

Status	Prior to	At termination/Current	Target	Gap
Maturity Level	2	3	5,6	3,4,5,6

Systems and Communication

System X was purchased without a thorough investigation of its suitability to Organization A's requirements. The design of System X is more compatible to the United States condition such as terms, terminology, function, and regulation. The result is that Organization A is facing a 'compatibility' problem, which ends-up with the conflict of system ownership between Academic Division and Information Services Division. The development of System X is time consuming and now enters its 7th year.

Status Prior to Project

Focus

Most IT systems in place are to coordinate information, and do not fully support the organizational needs for strategic advantage, such as used for decision making or managing the supply-chain. Legacy SIS is used to support the day-to-day operations such as storing, searching and retrieving student information.

Networking

Networking is limited to certain people (workgroups) in business units (Faculty/School/Academic Division/Support Services). For example, student information in Legacy SIS is only capable of electronic exchange within certain staff business units only. The information exchange inter-business unit is mostly done manually (by hard copy).

Current Status of the Project

Focus

IT focuses on cost reduction due to student data duplication and redundancy, streamlining communication organization-wide and improving the data quality. System X is currently integrated with Talis Library Management System (TLMS), which enables students to get automatic access to the library from the moment they are registered. The system's multiple-functionality is also capable of providing a basis for marketing strategies such as allowing enquiries to be identified and groups according to geographical location, educational background, interest in programmes/ modules etc. Apart from that, Organization A is currently being reorganized in phases to integrate disparate administrative and academic applications under one consistent access policy. This project is called the Identity Management Solution, which grants access to a wide range of applications used across Organization A including System X (student information), SAP (human resources), Agresso (finance), Talis (Library), and Athens (resources).

Networking

Most information is spread across Organization A through electronic mediums, where all business units are connected and the central IT/IS unit (ISD) provides communication services for all business units.

Target Status (the Readiness Level that Organizations Desire to be at in the Future)

Focus

The vision for the IT system is to extend the comprehensive services to potential students by enhancing the front and back ends of the system and extending the global market/geographic reach.

Networking

To facilitate Organization A's vision to extend global market reach, the networking thus must also be global in nature. The results of systems and communication analysis are shown in Table 2.

Process

The overall process of student information was not properly defined and documented, prior to the project and during the first three years of System X becoming 'live'; however, the vendor did not suggest any process change for System X to be successfully implemented. The management took a long time to react to this process

Table 2. IT Infrastructure: Systems and Communications

Status	Prior to	At termination/Current	Target	Gap
Maturity Level	2	3/4	5	4,5

problem, and the only efforts started during the third year of System X implementation, when a workshop was conducted to re-engineer the process.

Status Prior to Project

The processes were not properly defined; there is no assignment for overall responsibility and authority for performing the process and if it were done, in some cases – it was only done locally, in an ad-hoc manner and no training was provided for performing the role. The student information process is described as follows:

- *Programme Definition* – For each module the program definitions were defined by Programmes/Schools/Faculties and being kept and updated locally.
- *Student Enquiries* – Recording of the student enquiry information was done in various places with the information recorded in isolated systems, located at different location in the schools, faculties and Academic Division. Responses to enquiries were being dealt with locally, where the enquiry addressed the questions. Administration staff then batched and kept the enquiry details within their local systems.
- *Admissions* – Student applications are received electronically. Each application is then assessed, and if suitable, an offer is made. This process varies dramatically between Programmes/Schools/Faculties. The student replies and can then go through the interview process / open day, etc. UCAS applicants hold a firm offer (first choice) and an insurance offer. This proceeds to identifying accommodation and setting up the financial arrangements, i.e. identifying the fees. The process would then move on to registration. All of this process was done manually.
- *Registration & Enrolment* – Programmes/Schools/Faculties produce confirmation lists for registration & enrolment. Registration forms are completed with schools containing all the information related to personal details and modules. All arrangements related to the library, accommodation, and fees will be done during registration. This can lead to the double entry of data. Each of the business units such as the faculties, schools, library, and finance office have their own process in managing student information after this. Once registration is completed, the ID card is issued and the student becomes properly registered using all the facilities within Organization A.
- *Attendance* – Attendance registers are created by each programme. Attendance reports are submitted to Academic Division. Students can withdraw at any time and notification is by letter to the School Office or Accommodation. Facilities such as car parking will be cancelled and the fees revised.

- *Exams & Assessments* – Preparation of examination papers is handled by the Programmes. The central office that manages all exams will alert the Programmes/Schools/Faculties the pre-requisite conditions to be eligible to take the examination. Permission to appear in the examination is subject to settling the full amount of fees, library fines, etc. Academic Division will produce a list of candidates and exam scheduling for the examination and then send this information to Programmes/Schools/Faculties. All of the assessment marks such as coursework and examinations are entered onto a stand-alone system. Then the Programmes/Schools/Faculties will inform the Academic Division of the pass list for award purposes.
- *Graduation* – The award of degree to the student are after senate approval. The letter of attendance confirmation will be sent to students according to the address on record. Occasionally, the latest address is kept by Programme/Faculties/Schools and is not forwarded to the central administration. This results in central administration receiving incorrect data for attendance at the graduation ceremony.
- *Alumni* – All the latest information on graduate alumni students is kept in Academic Division.

The Criteria Relating to the Readiness Model in Level 1

Business processes were ad-hoc and with no documentation.

Current Status of the Project

The SIS student life cycle workshops were conducted to identify all the stages of the lifecycle requiring investigation in order to identify what needed to be mapped to System X's functionality. The workshop helped define the scope of the project as well as to identify all the business processes required to carry out each function. A well-defined business process includes standard descriptions and models of student life cycle from enquiry to alumni, and identifies responsibility and authority for performing the process. The student information process is described as follows:

- *Programme Definition* – For each module the programme definitions are defined by the Programmes/Schools/Faculties and being informed, kept, updated and centralized in System X. Each programme definition is submitted to an appropriate committee for approval. This includes the defining of assessment patterns, weightings, programme work and examinations. The details defined are all used during the student lifecycle.
- *Student enquiries* – Recording of the simple student enquiry information such as list of all available courses and additional information is currently

done centrally in System X. The enquiries are received via e-mail, telephone, letter, and are then manually recorded into System X. The response to enquiries is to be done either by automatic response or by email. The static and simple details of the enquiries will be automatically saved in the systems.

- *Admissions* – The entry application is received electronically and then on paper. Each application is then assessed and an offer made. This process is done centrally through System X. The student replies and can then go through the interview process / open day etc. UCAS applicants hold a firm offer (first choice) and an insurance offer. Then, system X automatically identifies the suitable accommodation with the applicants' criteria, timetabling, and setting up of the financial arrangements and identifying the fees. The process then moves on to registration.

- *Registration & Enrolment* – System X produces confirmation lists for registration & enrolment. Registration forms contain all the information related to personal details, accommodation, timetabling, fees, library, etc. Once registration is completed, the ID card is issued and the student becomes properly registered using all the facilities within Organization A.

- *Attendance* – Attendance registers are created for each programme via System X. Programmes/Schools/Faculties administrators key in reports to System X. Then, standard reports can be produced across programmes. Students can withdraw at any time and notify System X. Accommodation and other facilities such as car parking will be cancelled automatically and then the fee charge revised.

- *Exams & Assessments*. – The preparation of examination papers is handled by System X. Permission to be able to take the examination is subject to settling the full amount of fees, library fines, etc. System X will produce a list of candidates and the schedule for the examination can be checked on-line via System X. All of the assessment marks such as coursework and examinations are entered onto System X. Then pass lists will be automatically generated by the system.

- *Graduation* – The award of degree is given to the student after senate approval. The system keeps the information regarding the graduation ceremony, and passes it to the student for the attendance confirmation according to the latest address. The final lists of attendance will be produced and confirmed to the Programme/Faculties/Schools and students.

- *Alumni* – All the latest information of graduated students is kept within System X.

The Criteria Relating to the Readiness Model in Level 4

The business process is well defined with standard descriptions and models.

Target Status (the Readiness Level that Organizations Desire to be at in the Future)

Practices

The planning of System X enhancement targets for the future is currently underway:

- Record information about enquirers in a central database in order to provide statistical and marketing information for the organization.
- To build up a database of enquiry information that will enable the use of targeted marketing processes to improve the efficiency of the organization's recruitment activities.
- To build up a market view of the courses offered by the organization in respect of being able to identify areas of high and low interest, and hence to lead to the more effective planning of future products.
- To build up market views of the population interested in attending Organization A. This would cover both International and UK specific marketing activities.
- Enable the statistical breakdown of the data.
- To identify the processes required in order for functioning in the international market to be effective.
- It is expected to be fully replaced by the functionality the current System X will offer, which will then integrate the enquiry processes into the rest of the student life cycle.

The Criteria Relating To The Readiness Model In Level 5

Practices: The capability to set quality and measuring the products/services

The results of business process analysis are shown in Table 3.

Table 3. Business Process: Practices

Status	Prior to	At termination/Current	Target	Gap
Maturity Level	1	4	5	2,3,4,5

People

Skills

The System X project team does not have some skills in dealing with UNIX and YYY related issues. Apart from that, the System X project team also lacks the skills of project management prior to the System X project. The organization does not have good support from the vendor, no intensive training is provided by the vendor for the System X project team, and no business analyst position was present in the ISD prior to System X's implementation.

Status Prior to Project

Type of Skills

The IT/IS staff did not have adequate project management skills for System X system development, and Organization A had to appoint an additional programmer to deal with UNIX and YYY related issues. There was also a lack of business analytical skills to reengineer the student information process - users only acquired a basic IT/IS skill and knowledge to use basic IT functions and processes in the Legacy student information system. The users perform routine IT activities during the operation of the Legacy student information system.

Capability Building

The workgroups within the business units were identifying their own IT skill development, which was specific to their functions. There was very little sharing of information and knowledge between business units, and an insignificant amount of IT training provided to the System X users.

The Criteria Relating to the Readiness Model in Level 2

- Type of skill; basic and purely IT skills
- Capability building; team based/effort within workgroups in business unit.

Current Status of the Project

Type of Skills

The IT/IS staff gained project management (methodology, structured technique), technical (programming, networking) and analysis (process reengineer) skills and the experience needed for system development, and also obtained skills and confidence to install off-the-shelf readymade packages. Users are able to work and communicate between business units and monitor the student information process more efficiently. The required skills identified to operate System X are categorized in three stages:

- **Stage 1** – basic skills for the use and purpose of the system including essential searching skills and entering academic grades.
- **Stage 2** – skills to view and update the system including getting started on running standard reports, entering admissions decisions, decision making for direct applicants, altering and amending student records.
- **Stage 3** – concentrating on block processing, changing levels and programmes of study, exam board and academic standing.

Capability Building

There are centrally integrated training programmes within the organization conducted by the Education Development Unit (EDU) to facilitate the needs of staff in performing their work tasks. The training programmes conducted by EDU cover a wide–ranging area including IT/IS, management, communication, and leadership.

The Criteria Relating to the Readiness Model in Level 3/4

- Type of skills; considerable technical and project management skills
- Capability building; centrally integrated training program within organization.

Target Status (the Readiness Level that Organizations Desire to be at in the Future)

Type of Skills

Integration of key skills to be embedded and used within the future System X functionality to facilitate Organization A's vision to extend global market reach. The business units within the organization's reach have mutual understanding and appreciation of each other's roles and responsibility. Apart from managing student records, the users are also able to perform a broad range of tasks which are sometimes complex and non routine, such as dealing with multiple channels of customer responses in web portals, call centers, e-mails, voice-response to serve enquiries from customers and stakeholders. The users are also expected to demonstrate an analytical and systematic approach used for marketing strategy and planning. For example, the ability to segment groups of potential students according to demographics, gender, timing, behaviour, and lifestyle information, to identify the most likely potential students, and to improve marketing strategy and planning.

Table 4. People: Skills Analysis

Status	Prior to	At termination/Current	Target	Gap
Maturity Level	2	¾	5	3,4,5

Capability Building

It is expected, apart from training conducted by the center, that the staff also need to establish collaboration and knowledge sharing across the organization.

The Criteria Relating to the Readiness Model in Level 5

- Type of skills; gain cross-disciplinary experience to support supply chain activities.
- Capability building; knowledge sharing and management.

The results of skills analysis are shown in Table 4.

Roles and Responsibility of IT Staff

The development of the IT/IS unit has gradually changed over the past ten years. Prior to System X implementation, the Academic Information Service (AIS) was responsible for the organization's IT/IS development, and the Head of AIS was at middle management level. During 2002, there was a major restructuring of IT/IS units where the structure of AIS changed, and was renamed as the Information Service Division (ISD), where the role of ISD is to set IT strategy according to business needs. The Director of ISD was upgraded to the senior management group (Figure 6).

Status Prior to Project

Position of IT/IS Head

The post of IT Head was held by Head of AIS – positioned in the Support Service Group in the fourth layer of principal academic and administrative officer of the organization (Figure 6).

Roles

The major role is to provide technical assistance and advice to the wide-range of services across the organization including:

Figure 6. The position of the IT/IS head prior to the project

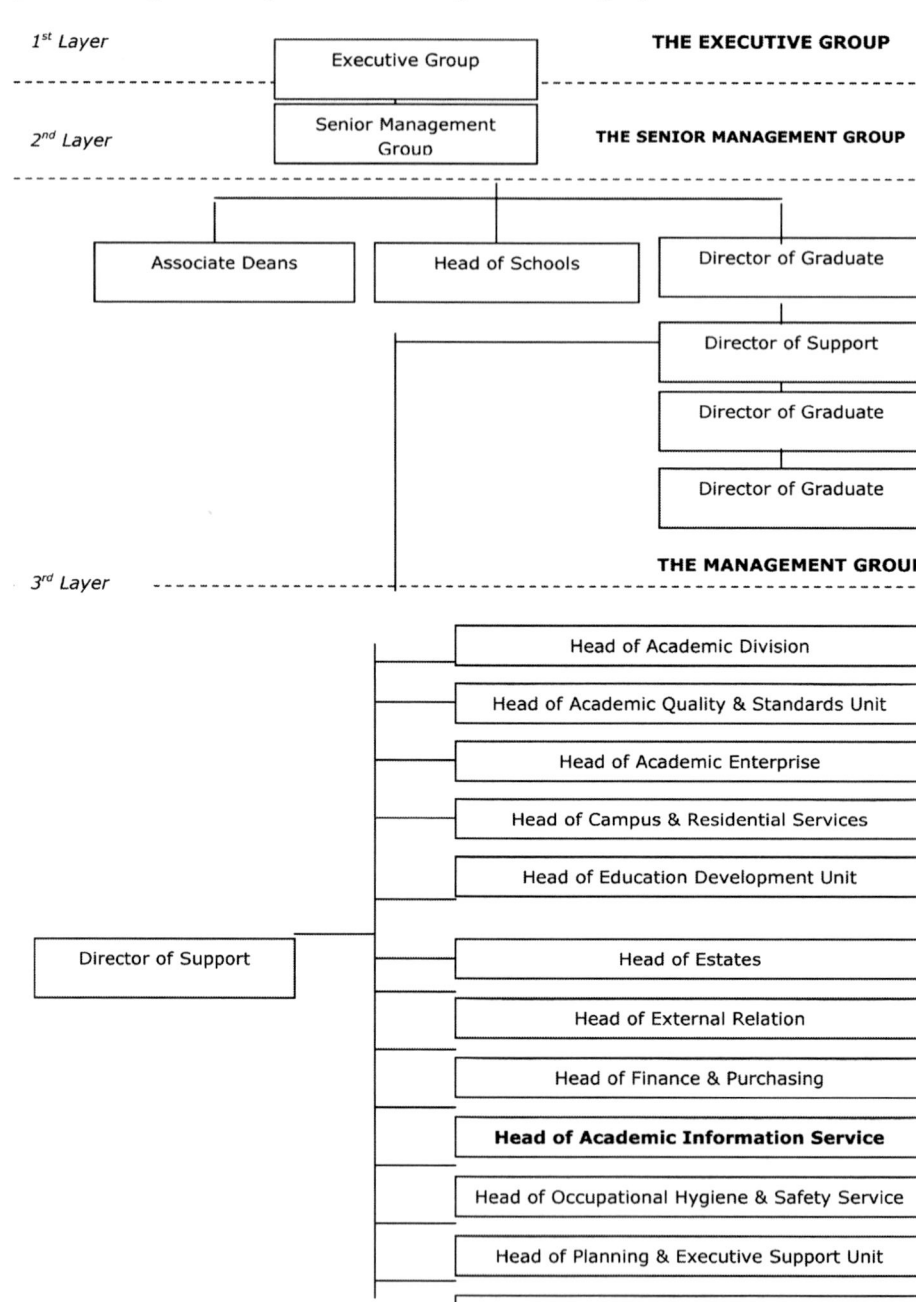

1. Technical support on a range of hardware across Organization A including personal computers, workstations, laptops, servers, printers, scanners, etc.
2. To supply Faculty/Schools/Programmes with specific software, that supports learning and teaching including applications and information systems, virtual learning, video-conferencing environments, electronic journals.
3. To provide a software copying service for Organization A within the constraints of legal licensing and purchasing agreements.
4. Technical support on a range of networking including the physical structure, whether cable or wireless, network servers, firewall, connections, switches, routers, internet access, etc.

The Criteria Relating to the Readiness Model in Level 3

- Position of IT/IS Head; Technical oriented manager
- Roles; technical support/purchasing and centralized IT/IS activities.

Current Status of the Project

Position of IT/IS Head

The post of IT/IS Head is held by Director of Information Service Division (ISD). The Director of ISD sits in the position of the senior management group in the second layer of principal academic and administrative officers in the organization (Figure 7).

Roles

General area of responsibility of Information Service Division (ISD) contained in the Policies and Procedure, Organization A, 2005:

Figure 7. The position of the IT/IS head at the current status of the project

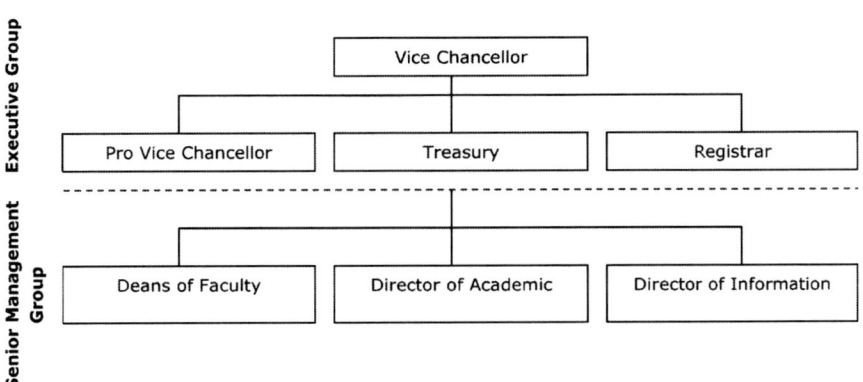

- All matters concerned with an information services strategy (embracing information and data management, business process analysis and information and communication technologies)
- The derivation and maintenance of an information services strategy aligned to the objectives of Organization A
- The derivation and implementation of an organization-wide policy for IT investment in both the ICT technologies and information resources (all media types).

The Criteria Relating to the Readiness Model in Level 5

- Position of IT/IS Head: IT Manager with senior management status
- Roles; IT/IS strategy.

Target Status (the Readiness Level that Organizations Desire to be at in the Future)

Position of IT/IS Head

The management hierarchy of Organization A is expected to remain the same in the future.

Roles

The role of ISD within Organization A is expected to remain the same in the future.

The Criteria Relating to the Readiness Model in Level 5

- Position of IT/IS Head: IT Manager with senior management status
- Roles; IT/IS strategy.

The results of roles and responsibility of IT staff analysis are shown in Table 5.

Table 5. People: Roles and Responsibility of IT Staff

Status	Prior to	At termination/Current	Target	Gap
Maturity Level	3	5	5	4,5

User Involvement

The extent of user involvement in the IT system development improved prior to the project, and at the termination/current time, and future undertakings.

Status Prior to Project

User Involvement

Initially, the shortcomings of Legacy SIS have been notified by the users in Academic Division (AD). For example, there has been a serious problem regarding updating student information due to the isolation of the central database. This matter has been raised by the Head of AD during the support service group monthly meetings. The Director of Support Service then brought this issue to the attention of the management group in their regular meeting. The AIS were assigned to look at this problem and prepare a report and recommendation to the management group, and strongly recommended that Legacy SIS needed to be replaced with a central student database. The senior management group gave the green light for the AIS to pursue new system development to replace the old system, with an ad-hoc user committee being formed within AIS to identify the basic needs and requirements of the new system.

The Criteria Relating to the Readiness Model in Level 2

• Individual consultation

Current Status of the Project

User Involvement

At the initial stage, a small user ad-hoc committee that formed within AIS identified the basic needs and requirements of the new system. However, there was a serious problem relating to the functionality of the new System X which did not fit to Organization A's needs. Under the Stabilization Project, a dedicated committee for System X development was formed which included the focus user group called the SIS User Council (SISUC). The SISUC acts as a forum for SIS Senior Contacts and representatives of central administrative units to review how System X is currently used to support their work, to review proposed SIS developments, and to recommend future developments. The membership of SISPUC comprises faculty representatives, school representatives, support services representatives and a System X project team.

The Criteria Relating to the Readiness Model in Level 3

- Focus group consultation (purchasing & centralizing IT activities).

Target Status (the Readiness Level that Organizations Desire to be at in the Future)

User Involvement

The user involvement in IT/IS projects is expected to improve by participating in not just problem definition and systems requirements, but extending beyond to IT/IS strategy, approval and testing the software. This is due to future planning System X's development towards creating a more interactive environment between the internal users and customers.

The Criteria that Relates to the Readiness Model in Level 5/6

- Ad-hoc/permanent user group in IT project team.

The results of user involvement analysis are shown in Table 6.

WORKING ENVIRONMENT

Organization Behaviour

Prior to the project, each of the business units within Organization A had their own separate process in managing student information. The design of System X from a technical point of view then persuaded the users to adapt to it. Initially, the success of System X was measured in terms of IT functionality rather than the impact on the organization's operation. This has caused a problem in suiting the organization's needs. The benefits of System X are measured quantitatively such as saving produc-

Table 6. People: User Involvement Analysis

Status	Prior to	At termination/Current	Target	Gap
Maturity Level	2	3	5/6	3,4,5,6

tion costs and time. After three years of System X becoming 'live', the system was adjusted according to the process.

Status Prior to Project

Characteristics

Each of the business units, i.e. Schools/Faculties/Academic Division/Support Services had their own way of managing student information. For example, most of the Schools/Faculties preferred the manual handling of student information, i.e. files and hardcopy. The user's attitude in various business units was at a different level. For example, users at Academic Division (AD) and Library Services perceived IT as a vital element in performing their daily work tasks, whereas other business units perceived IT as one of many options to perform their work task. Less interaction and a lack of student information sharing particularly between Schools/Faculties/ Programmes and Academic Division (AD) created problems, such as a lack of a single source of student information creating a data crisis through a lack of information, which was accurate and up-to-date.

The Criteria Relating to the Readiness Model in Level 1

- Non data sharing culture exists.
- Each individual has own separate processes.

Current Status of the Project

Characteristics

The selection of System X was based on the track record of the vendor supplying and installing the system world-wide, but without considering it's suitability to the organization's needs and requirements. For example, no thorough research was done to assess what was going to change in the current organization's work environment when System X was introduced. As a result, System X causes many conflicts, such as failing to deliver the expected results, and it therefore needs major customization efforts to suit the organization's work environment. Organization A is slowly coping with the new ways of managing student information throughout its lifecycle. The user sees the benefits of System X as more to speed up their daily work tasks (efficiency) rather than the impact made on the business (effectiveness). For example, the issues raised during management meetings were how to improve the operation of the system. Apart from that, System X was also perceived as one way of reducing many printing costs, such as the minimization of prospectus publication by introduction of an e-prospectus.

The Criteria Relating to the Readiness Model in Level 2

- Technology-led.
- Design the system from technical point of view.
- IT success is measured in IT terms rather than the impact made on the business.
- Monitoring expenditure.

Target Status (the Readiness Level that Organizations Desire to be at in the Future)

Characteristics

The improvement of a future System X that provides better services to the potential students by allowing direct entry of information by remote enquiries, or data to be entered by administrative staff on behalf of the enquirer. The registered student will also be given permission to update their information, such as term time address, contact number, etc. Apart from that, the users also use System X as a marketing strategy by analyzing the source of all enquiries in statistical data such as location, interest, favorite course, gender, etc.

The Criteria Relating to the Readiness Model in Level 5

- IT/IS used to provide better products/services to customer.

The results of organizational behaviour analysis are shown in Table 7.

IT Policy

The centralized ISD unit controls all IT/IS activities across the organization. However, the business unit is allowed to purchase/develop their own IT/IS within ISD.

Table 7. Work Environment: Organizational Behaviour Analysis

Status	Prior to	At termination/Current	Target	Gap
Maturity Level	1	2	5	2,3,4,5

42

Status Prior to Project

IT/IS Activities Control

Organization-wide network infrastructure, ICT system (including hardware and software).

The Criteria Relating to the Readiness Model in Level 3

- Centralized IT infrastructure and application.

Current Status of the Project

IT/IS Activities Control

ISD is authorized to manage the organization-wide network infrastructure, ICT system (including hardware and software)

- The business units; Schools/Faculties/Support Services are authorized to maintain the following infrastructure:
  - Machine specific hardware problems (e.g., drives, mouse, keyboards)
  - Monitors
  - Local and network printer faults
  - Peripheral equipment (e.g. scanners, external drives)
  - Local software faults (e.g. MS Office problems)
  - Operating systems
  - Anti-virus problems
  - Data and document backups
- Most faculties and research institutes have some research project collaboration with industries that required purchasing specific hardware/software. In this case, the respected faculties and research institutes have to inform the ISD prior to purchasing.
- There is a procedure (policy) to be followed when the School/Faculty/Support Service requires specific software/hardware by informing ISD about:
  - The purpose of the software/hardware
  - When the software/hardware is required
  - The expected number of users
  - Any accessibility restriction
  - The contact details of the Schools/Faculty/Support Service staff supporting the software

- Then, ISD will inform Schools/Faculty/Support Services about:
  - The decision on the software/hardware application within one month
  - The date when the software/hardware is ready for testing.
  - Restriction of access to the software/hardware, and ensure back up and restore functions as needed
- ISD measure their service to Schools/Faculty/Support Services in two ways:
  - Quantitative – they record how many requests are processed on a monthly basis
  - Qualitative – through Customer Satisfaction Surveys.

The Criteria Relating to the Readiness Model in Level 4

- Centralized IT infrastructure, and beginning to implement decentralized application (a policy), with intervention from the IT Department.

Target Status (the Readiness Level that Organizations Desire to be at in the Future)

IT/IS Activities Control

IT infrastructure organization-wide is expected to remain under the management of ISD, whereas the Schools/Faculties/Support Service have the full authority to purchase their own applications without intervention from ISD.

The Criteria Relating to the Readiness Model in Level 5

Centralized IT infrastructure and establishment of decentralized applications (with no intervention from the IT Department).

The results of IT Policy analysis are shown in Table 8.

Leadership

Organization A acquired integrated communication planning that has a specific type, target, medium, etc., across the organization. The senior management is involved

Table 8. Work Environment: IT Policy Analysis

Status	Prior to	At termination/Current	Target	Gap
Maturity Level	3	4	5	4,5

after System X is in place, and the ISD is a centralized communication group across the whole organization.

Status Prior to Project

Communication

There is a procedure and policy organization-wide regarding communication. However, the medium of communication varies and is mostly by publishing in newsletters and email circulation. No internal communication strategy exists within the business unit itself, and mostly on an ad-hoc basis.

Participation

There is senior management participation in the IT/IS project, however the participation is through the agenda's discussion and notification during regular senior management meetings, and more on general matters such as project status (a milestone comparing plans to their realization).

The Criteria Relating to the Readiness Model in Level 3-2

- Communication; organization-wide policy and standards for communication.
- Participation; ad-hoc participation.

Current Status of the Project

Communication

The Performance and Communications Unit under ISD holds the responsibility for the communication strategy and planning throughout the organization. ISD's communication objectives are:

- To improve customer relations through improved communications, better feedback mechanism and opportunities for consultation.
- Develop plans, policies and procedures that will support the improvement of communications.
- Improve channels, tools, information resources and ICT training that supports and enables communication flows.
- Eliminate barriers to effective communication and, where feasible, tackle information overload that relates to the use of communication technologies.

Table 9. Communications cascade framework

PERIOD	Daily	Weekly	Monthly	Quarterly	Annually
TYPE	Message of the Day	ISD News	ISD Updates	Director's Personal Email	ISD Staff View Univ. Staff View
TARGET	Organization members	All ISD Staff	Organization members	Individual ISD Staff	ISD Staff
AUTHOR	Any organization related sources	Any ISD Staff member	ISD Editor/ISD member	Director	ISD Editor
EDITORS	P & C Staff	Training Admin-istrator	ISD Editor/Head of P & C	N/A	Head of P & C
FACILITATORS	ISD Editor	Head of P & C	Head of P & C	Head of P & C	Head of P & C
FORMAT	Electronic-log in launch	Elec./Printed Newsletter	Elec./Printed Newsletter	Personal Email	Questionnaire
CHANNEL	ISD Web	E-mail/Notice Board	ISD Websites/Notice Boards/leaflet racks	E-mail	ISD Websites
SCOPE	Any matter of wider interest	Operational, service-wide	Specific and general, esp. technical	Progress, barriers, aspirations, motiva-tion and culture	Seek opinion
ARCHIVE	Database	ISD Intranet	ISD Websites	E-mail server	ISD Websites

The selection of communication channels to be used by ISD depends on the content of the information, speed with which it needs to be delivered, and the intended audience. The current communication channels used by ISD are:

- ICT Software
- Directories
- Websites
- Intranets
- Messages of the Day

The overall organization communication strategy is set in the Communications Cascade Framework 1, which is summarized in Table 9. All the business units also need to produce their own communication strategy based on local activities (Table 10).

Participation

The participation of senior management in Organization A with the System X project is through membership in the project executive committee group namely, the Student Information System Project Executive Group (SISPEG). The membership of SISPEG comprises various representatives from Schools/Faculties/Support Service management which comprises of:

- Pro-Vice-Chancellor Institutional and Student Services (Chair)
- Director of External Relations

Table 10. Local activities communication strategy

PERIOD	Monthly	Monthly
TYPE	Internal Communication	Unit Activity Report
TARGET	Individual Staff/Group Team	Individual Staff/Group Team
AUTHOR	Line Managers/Team Leaders	Line Managers/Team Leader
EDITORS	N/A	N/A
FACILITATORS	Head of Unit	Head of Unit
FORMAT	Varies (primarily verbal & email)	Varies
CHANNEL	Varies to meet local needs, especially part-time and weekend staff	Varies
SCOPE	Local implications for service wide issues	Unit Updates, achievements, activities/plans
ARCHIVE	Any record agreed locally	Any record agreed locally

- Associate Director Information Systems, ISD
- Director of ISD
- Chief Accountant, Finance Division
- Project Delivery Manager, Information Systems, ISD
- Academic Registrar
- Chair of the User Council
- School and Faculty Representative
- Registrar and Secretary
- SIS Service Delivery Manager
- SIS Project Administrator (Secretary).

The SISPEG has overall responsibility for the funding, direction and progress status of the System X project, and all other committees associated with the project report to it. SISPEG meets every two months, and the committees like SISPEG only exist for large and high scale IT projects.

The Criteria Relating to the Readiness Model in Level 5-3

- Communication; well documented and integrated communication planning for overall organization structure.
- Participation; Participation in large scale and high cost IT projects.

Target Status (the Readiness Level that Organizations Desire to be at in the Future)

Communication

ISD aims to improve the effectiveness of the communication plan by setting a few key performance indicators and performing a series of customer feedback, both internally and externally:

- An annual ISD staff survey focused specifically on communications at a general level and individual level.
- The ISD Performance and Communication Unit will collate results of progress including results of any relevant key performance indicators which are currently under development, and will then produce an annual summary.

Participation

System X is expanding its scope by covering postgraduate information, and the project is on its way and expected to be continued in two years' time. The senior management was also involved in this project as a member of the Student Information Service Postgraduate Project Board (SISPGPB). The role of SISPGPB has a different role to SISPEG. These are as follows:

- Identify major risks and agree action plans to manage these.
- Monitor project plans and timescales produced by the project team against agreed project priorities.
- Act as the escalation point for issues which cannot be resolved by the project team.
- Ensure the project delivers value for money.

The Criteria Relating to the Readiness Model in Level 6/4

- Communication; continuous improvement of the communication process
  Participation; participation on all IT/IS activities.

The results of Leadership analysis are shown in Table 11.

Table 11. Work Environment: Leadership Analysis

Status	Prior to	At termination/Current	Target	Gap
Maturity Level	3-2	5-3	6-4	Improvement

SUMMARY AND FINDINGS

It is identified that the overall process of student information was not properly defined and documented prior to the project, and during the first three years of System X becoming 'live' (level 1-process). Major work was needed to improve processes in order to maximize the potential of System X. For example, the student information process scopes needed to be identified and standardized organization A-wide, with a clear definition of roles and responsibility (level 2, 3, 4-process). In the case of System X, the management took a long time to react to this process improvement, and the only effort started during the third year of System X implementation, when a workshop was conducted to re-engineer the process. The vendor did not suggest any process changes that were suited to System X.

Prior to the project, each of the business units such as the Faculties, Schools, Academic Division, the Library, etc. had their own separate processes in managing student information. The user's attitudes in various business units were at a different level. For example, users at Academic Division (AD) and Library Services perceived IT as a vital element in performing their daily work tasks, whereas other business units perceived IT as one of many options for performing their work tasks (level 1 – organization behaviour). A lot of effort was directed towards the improvement of the work culture across Organization A such as to change and synchronize the attitude and perception of the users towards the potential of IT/IS for improving their work tasks (level 2,3,4,5 – organization behaviour).

Organization A has an excellent structure of the IT unit. However, due to the major restructuring of IT departments in 2002, where the structure of IT units changed (formerly known as Academic Information Service and renamed as Information Service Division), this influenced the implementation and planning of System X (level 3-roles and responsibility of IT Staff). Prior to IT unit restructuring, the post of IT/IS Head was held by the Head of Academic Information Services (AIS) and sat in the Support Service Group in the fourth layer of principal academic and administrative officers of the organization. This was to provide technical assistance and advice to the wide-ranging services across the organization. After restructuring, The post of IT/IS Head was held by Director of Information Service Division (ISD) and was a member of a senior management group in the second layer of principal academic and administrative officers of the organization. This was to deal with all matters concerned with information services strategy (levels 4,5 - roles and responsibility of IT Staff).

The IT unit centralized all IT/IS activities across the organization since IT/IS was introduced in Organization A (level 3 – IT/IS policy). This policy remains the same, but more flexibility has been given to the faculties/schools and research institutes to purchase/develop their own IT/IS due to the collaboration work with industries and others universities (level 4, 5 – IT policy).

Prior to the project, the IT plan was more focused on work task requirements (level 2 – top management perception). In another development a few years later, in 2004, the senior management launched *'The Strategic Framework 2005-2015'* containing Organization A's long term mission, goals and priorities for the more immediate action, including IT/IS investment toward global competition (level 3,4,5,6 – top management perception).

Prior to the project, there was a procedure and policy organization-wide regarding communication. However, the medium of communication varied and was mainly by publishing information in the newsletter and by email circulation (level 3 – leadership). Currently, Organization A has acquired integrated communication planning that has a specific type, target, medium etc., across Organization A (level 4, 5 –leadership).

Figure 8. Summary of assessment

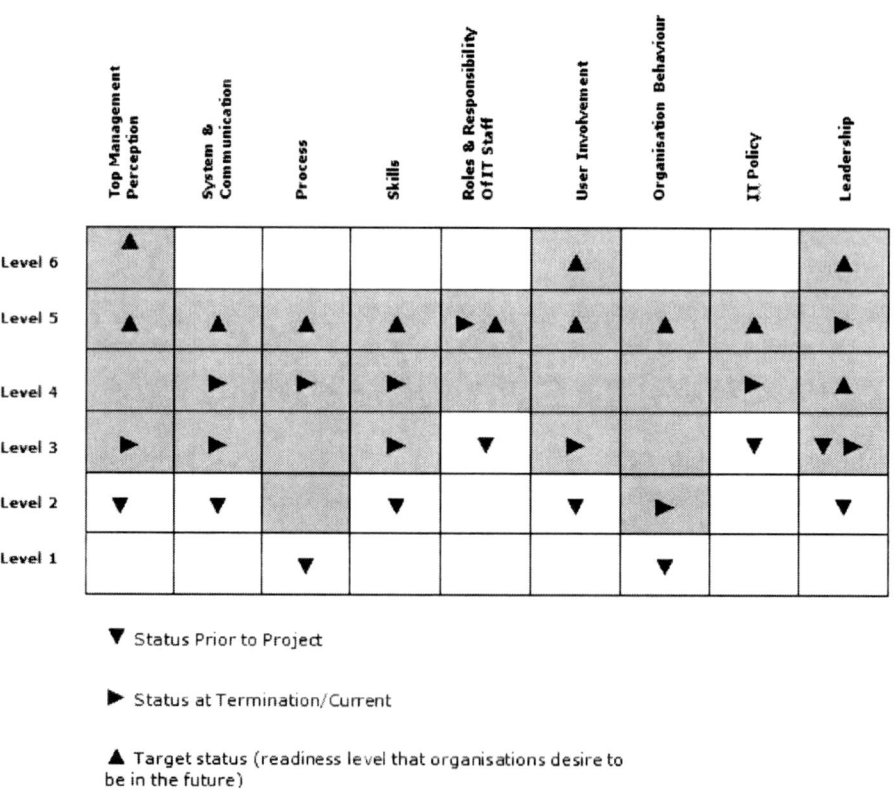

	Top Management Perception	System & Communication	Process	Skills	Roles & Responsibility Of IT Staff	User Involvement	Organisation Behaviour	IT Policy	Leadership
Level 6	▲					▲			▲
Level 5	▲	▲	▲	▲	►▲	▲	▲	▲	►
Level 4		►	►	►				►	▲
Level 3	►	►		►	▼	►		▼	▼►
Level 2	▼	▼		▼		▼	►		▼
Level 1			▼				▼		

▼ Status Prior to Project

► Status at Termination/Current

▲ Target status (readiness level that organisations desire to be in the future)

Gap

Prior to the project, the user got involved in the IT/IS project on an ad-hoc basis (level 2 – user involvement). During System X implementation, wider user representation was involved across all business units in the stabilization project, the dedicated committee for System X's development. A lot of improvements should be done to get more user participation in IT/IS projects, such as forming a permanent user group for IT/IS projects and looking for an IT/IS and business strategy (level 3,4,5,6 – user involvement).

A lot of work is needed for Organization A to improve the maturity gap in achieving their business objectives, particularly in the area of process, user involvement and organizational behaviour. The summary of the assessment of System X implementation according to the readiness model can be found in Figure 8.

CASE STUDY ADOPTED FROM

Saleh, H. (2007), "Measuring Organisational Readiness Prior to IT/IS investments", PhD Thesis, School of the Built Environment, The University of Salford.

Chapter 3
Improving the Tendering Process:
A Construction Organization Perspective

Hafez Salleh
University of Malaya, Malaysia

EXECUTIVE SUMMARY

Construction sector is unique in a way because more than 90% companies are Small and Medium Sized Enterprise (SME). This chapter presents a case study of a construction company with past and current projects valued up to £15 million, and has completed a number of construction projects both in private and the public sectors including housing, commerce, leisure, health, education, retail, et cetera. The company operates out of multiple locations and decided to improve the tendering process using a new IT system. On applying the IT readiness model it was found that almost all the attributes identified were not at lower level 1. The practice achieved maturity in three areas – skills and leadership are identified at level 5, and roles & responsibility were identified at the top of the level.

BACKGROUND AND HISTORY

Organization C is a construction company with past and current projects valued up to £15 million, and has completed a number of construction projects both in private and the public sectors including housing, commerce, leisure, health, education, retail, etc. Organization C's headquarters are located in England with a turnover reaching

DOI: 10.4018/978-1-61350-311-9.ch003

Figure 1. General structure of The Main Building Group

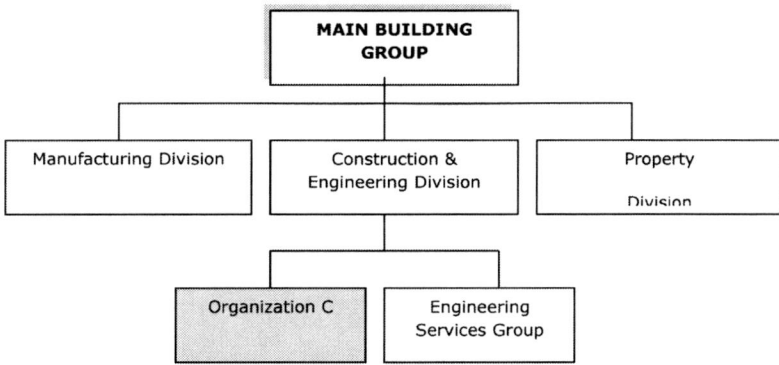

£370 million. Organization C is one division of its main building company. The general structure of the main building group is shown in Figure 1, and consists of one head office and six branches as shown in Figure 2.

Timeline

2002

- There were many limitations of the previous system in managing the practice. Sub-contract tendering was reported from the staff.
- A key issue is interfacing between two different IT systems, between two units in the organization, who actually managed and operated the process, namely, the estimator, and the print room.

Figure 2. Organization C's offices

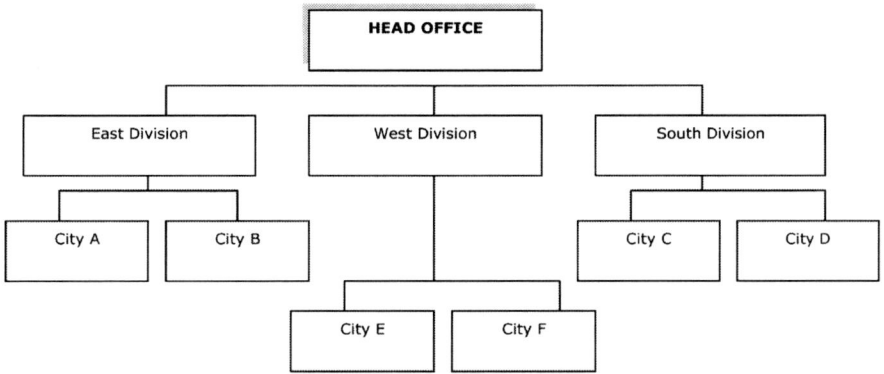

- There was no single documentation to define and describe the sub-contracting tender process.
- Every estimator had a different perception and definition of the sub-contracting tender process and therefore cost, time and quality mainly relied on the estimator capabilities in performing their job.
- Because of this issue, a lot of problems arose:
  - Duplication of input
  - Errors/manual process of cross-referencing
  - Searching for information
  - Tracking and controlling revisions
  - Administering related procedures.
- In 2002, the management practice decided to seek a new IT system that solved or minimized these problems.

2003

- There was a series of meetings and presentations between the management practice representatives and vendors that were looking for a suitable system to improve the sub-contract tender enquiry process.
- A system proposed by Vendor C was selected and called System Z.
- Vendor C offers a range of packages for the construction and property related industries.
- System Z is a document management system (DMS), and provides collaboration solution that is ideal to control and manage the creation of drawings and associated documents throughout sub-contract enquiry processes.
- Vendor C provides a high commitment to System Z's development;
  - By providing the package,
  - System requirements definition – functional and non-functional (providing an assistance to in-house expertise).
  - System analysis – improved sub-contract tendering processes (working with in-house expertise).
  - System design – developing detailed specifications and hardware/software recommendations.
  - System development – providing assistance to in-house expertise.
  - System implementation and support – including training (working with in-house expertise).
  - System maintenance – providing assistance to in-house expertise.
- In-house expertise has the capability to develop and maintain IT/IS such as programming, analysis, being able to install and manage off-the-shelf, ready-made packages, etc.

Figure 3. Vendor C's proposed stages of System Z implementation

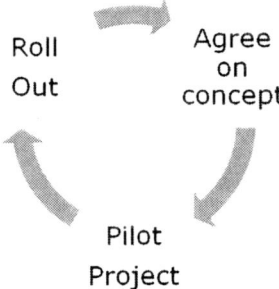

- In the early stages, Vendor C assisted the in-house expertise to analyze the existing business processes and recommend the solution.
- Vendor C also provides a dedicated on-site Project Manager to coordinate the process.
- Vendor C's proposed stages of System Z implementation is shown in Figure 3.

- In June 2003, Vendor C and the management practice had agreed on the concept of the proposed System Z as follows:
  ○ Business analysis
  ○ Functional and non-functional specification
  ○ Development
  ○ License profile
  ○ Training
  ○ Support
- At that time, there were several training programmes conducted within the practice. However, the training was more focused within the business unit. Therefore, for System Z development the management practice plan was to conduct more centralized training.
- In most cases, user involvement in IT/IS development was on a group and ad-hoc basis, and only involved high cost IT/IS projects.

2005

- In November 2005, the pilot project was done.
- In December 2005, the trial version of System Z was live for three months for reviews.

2006

- User representatives came from relevant groups of people. For System Z, all the estimators were involved in prioritizing the system's requirements.
- Finally, after a long run of system design and planning, System Z became fully live in March 2006.
- System Z enabled the whole sub-contracting tender process to be repeatable for the new sub-contracting documents, which allows organizations to repeat successful practices developed on earlier projects.
- Policies for managing sub-contract quotations and procedures to implement those policies were established.
- The hardware and software requirements for System Z are Windows 2000/2003, run on a UNIX platform, and a YYY database.
- Currently, each of Organization C's offices has managed their own system according to their project needs.
- Figure 4 illustrates the timeline for System Z's development.

The practice expected benefits of System Z are to:

- Reduce duplication, i.e. cross referencing
- Improve document searching
- Reduce reliance on hard copy distribution
- Reduce reliance on manual version control

Figure 4. System Z's development timeline

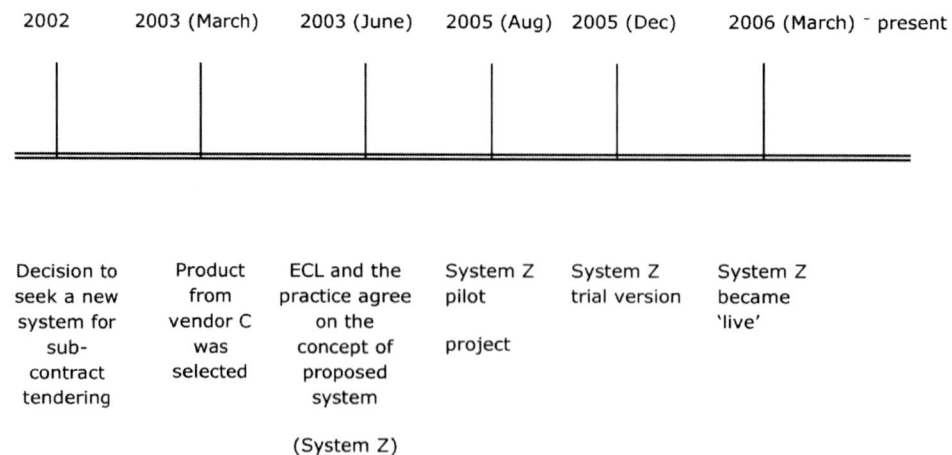

- Administer standard practice/procedures
- Integrate with existing systems (CDS/Oce)

ANALYSIS AND DISCUSSION

The following section addresses each attribute of the four domains of the readiness model within the context of Organization C (Head office, York).

IT INFRASTRUCTURE

Top Management Perceptions

There was high awareness towards the potential of IT for the practice among the senior management. The Managing Director and IT Director were actively involved in the IT research activities between industry and academia. The IT Director was on the board of directors and has great influence on the direction of the practice IT/IS investment and most of the IT/IS project focused on improving and smoothing the business process (efficiency). There was an intention (IT plan) to upgrade/ enhance the existing system for competitive advantage (effectiveness). The *status prior to project* refers to the period prior to 2006 in which the mechanism of the sub-contracting tender process operated between being 100% paper based, and 50% manual - 50% electronic. The *current status of the project* refers to the period when System Z became 'live' in 2006, where 80% of their sub-contracting tender process operated electronically. *Target status* refers to the readiness level that organizations desire for 100% electronic operation of their sub-contracting tender processes in the future.

Status Prior to Project

Drivers

IT investment was to support processes that were essential for the company to run its business; this process included producing the tender document for sub-contractor tender. The applications to support the process were the most regularly used in the construction industry, such as CATO estimating software to support producing tender documents and AutoCAD design software to support producing drawings and specifications. Most of the IT investment was on the same scale as other similar companies in the industry.

- Database system – in house, similar to SAP and accounting systems, i.e. led-gers, payment to subcontractors, purchasing, assets, cost control valuations to run the business.
- Oce – tender documents indexed
- CDS – Bill of Quantities indexed
- Office suite
- Printing hardware
- Shared drive for storage

System Requirements Definition

In most cases, the in-house experts prepared the initial brief of the functional and non-functional requirements. The vendor makes further detail of functional and non-functional requirements.

The Criteria Relating to the Readiness Model in Level 2

- Drivers; work task requirements
- System requirements definition; partially from in-house and mostly from vendor.

Current Status of the Project

Drivers

The problems encountered using the previous systems in managing sub-contract process were:

- The high cost of reproduction of large documentation for producing tender documents, including bills of quantities, specification, preliminaries, and drawings.
- Time taken for compiling those documents.
- Large space needed to allocate filing cabinets to store the volume of large sized documents.
- Relatively high postage costs to send tender documents.
- Sometimes, there was a difference in the tender document received by sub-contractors, due to human error during the compiling process. This led to the issue of sub-contractors pricing the tender on the basis of incorrect information, and resulted in disputes during the tender and construction process.

To solve the problems, the management decided to switch to System Z in order to make the sub-contract tendering process more efficient. By using System Z, the tender document could be kept in a centralized electronic database management system (EDMS) and produced in CD format:

- Reducing hard copy distribution
- The only cost related to reproduction is to make a copy from the original CD
- Efficient document sharing
- Save time due to less time taken to reproduce the tender document
- No large space needed to store the tender documents
- The sub-contractors are guaranteed to receive the same information contained on CD.

System Requirements Definition

The vendor provided assistance to in-house expertise in systems requirements definition (functional and non-functional).

The Criteria Relating to the Readiness Model in Level 3

- Drivers; cost reduction, organization communication
- System requirements definition; mostly in-house with intervention from vendor.

Target Status (the Readiness Level that Organizations Desire to be at in the Future)

Drivers

There was great demand from the industry to change the traditional practices to the automation process. The existing System Z was half electronic, whereby the sub-contractor had to send the pricing tender by post and the tender analysis was also done manually; and selection of the sub-contractor was also limited to those who had a history of working experience with the practice. Thus, the practice plan was to upgrade System Z by extending the market and geographic reach of the tenderers (sub-contractors) and fully automate the whole process of sub-contract tendering by introducing an e-tendering system. The e-tendering also enabled the subcontractors to interact with the system, and the practice, getting information back in a standard format. However, the practice at that time did not have the intention to extend their business overseas.

System Requirements Definition

In order to implement e-tendering, the practice was expected to co-operate with sub-contractors and vendors to determine the needs and requirements.

The Criteria Relating to the Readiness Model in Level 5

*   Drivers; managing supply chain
*   System requirements definition; mainly in-house with intervention from supply chain partners and vendors.

The results of top management perception analysis are shown in Table 1.

System and Communication

System Z was developed with investigation of its suitability to the practice requirements to improve the sub-contract tendering process. A pilot project and trial version included in the development phase allowed for the practice to identify an error and rectify it prior to actual implementation. Vendor C provided good support to in-house expertise during System Z's development and System Z successfully became live after three years of planning and development.

Status Prior to Project

Focus

IT/IS development in the practice was intended to support internal operational processes. For example, QS used CATO program CAD measures (ECL product) to measure quantities direct from drawings (digitizer). The designer used AutoCAD for design, the Oce System for scanning and printing and CDS for indexing.

Network Communication

In the early days, the entire PC was on a standalone basis. Then it was upgraded to enable the communication between workgroups in a business unit. Everyone used

Table 1. IT Infrastructure: Top Management Perception Analysis

Status	Prior to	At termination/Current	Target	Gap
Maturity Level	2	3	5	3,4,5

self IT services, such as printing and scanning; and document exchange was mostly on a manual basis.

The Criteria Relating to the Readiness Model in Level 2

- Focus; co-ordinate information activities
- Network communication; networks within workgroup in business unit.

Current Status of the Project

Focus

IT/IS development devoted to improve the tender documentation process and to managing the selection of a successful sub-contractor in the tender.

Network Communication

Every PC was connected to a network and shared some common IT services such as printers, scanners and servers, organization-wide. The practice installed the Windows Server 2003 operating system and Microsoft Exchange 2003, replacing the Windows NT4 Server operating system and Microsoft Exchange 5.5, to manage and control all messaging from head office. The practice also developed the Intranet for publishing tools that allowed multiple content publishers to manage their department information and helped organize existing information into a site that was easy to navigate, was well structured, and allowed for future expansion. The information in the Intranet also allowed users to not only work with text and images but also to create feedback forms, document libraries, team pages, etc.

The Criteria Relating to the Readiness Model in Level 4

- Focus; Improve decision making
- Network communication; network used organization-wide

Target Status (the Readiness Level that Organizations Desire to be at in the Future)

Focus

The practice of IT/S development was expected to extend its function to support managing the information supply throughout industry.

Table 2. IT Infrastructure: Systems and Communication Analysis

Status	Prior to	At termination/Current	Target	Gap
Maturity Level	2	4	5	3,4,5

Network Communication

It was expected that the future system of e-tendering for the sub-contracting process, was to be fully managed electronically, and so, subcontractors would be able to view and get a sub-contracting package irrespective of their location. The practice also planned to use Broadband wherever it was available, GPRS access for a mobile workforce, wireless networking on site, thin client technology and the use of PDAs.

The Criteria Relating to the Readiness Model in Level 5

- Focus; manage supply-chain
- Network communication; network used to improve supply-chain performance.

The results of the systems and communication analysis are shown in Table 2.

PROCESS

Practices

The sub-contract tendering process maturity was managed well and followed the expected sequence; *manually, half manually and full electronic* over a period of time, and therefore the transition was smooth. The vendor played an important role in the business process improvement during planning and development of the System 'Z' stage; the trial version of System Z for three months prior to real live implementation helped the practice absorb the new practice, and allowed time for alteration.

Status Prior to Project

Practice

- 1st phase (late 80's) – *100% paper based*
  - Main tender document received in hardcopy
  - Estimator determined which pages were to be photocopied

- ◦ Then taken out from original bills of quantities for photocopy
- ◦ Printed and compiled for dispatch as sub-contract package (postal)
- ◦ Received package back from sub-contractors via post and the estimator evaluated manually
- 2nd phase (late 90's - 2006) – *50% manual and 50% electronic*
  - ◦ Main tender document received in hardcopy
  - ◦ Estimator determined which pages were to be scanned
  - ◦ Then taken out from original bills of quantities for scanning (Oce Scanning and Printing System) and indexed with CDS (Index System)
  - ◦ Printed and compiled as sub-contract package for dispatch (postal)
  - ◦ Received package back from the sub-contractors via post and the estimator evaluated manually.

Most of the process was standard practice within the company, however there was no single document to define and describe the process available. Therefore, the company experienced wide variations in cost, delivery times and quality targets. This was due to the fact that every estimator had a different perception and definition of the process, and all of these elements (cost, time and quality) rely on the estimator's capabilities in performing their job. The management of the new sub-contract quotation document was based on experience with similar projects. The problems that occurred during the process were solved in an ad-hoc manner and depended on the previous experience of the estimator in handling similar situations. Thus, the success of the work process depended entirely on having an exceptional estimator, and the whole process would have been affected if the estimators had left the company. The quality of information was not guaranteed, due to a tendency of duplication of input, or error due to removal of pages from the bills of quantities, scanning, and indexing. There was no assigned responsibility to manage the process, and there were problems on storing, tracking, and controlling documents.

The Criteria Relating to the Readiness Model in Level 2

Practice: the business processes scope was identified and improved as the work progressed.

Current Status of the Project

Practice

- 3rd phase (2006) – *80% electronic*
  - ◦ Main tender document received in hardcopy

○   Estimator determined which pages were to be scanned
○   Then taken out from the original bills of quantities for scanning (Oce Scanning and Printing System) and indexed with CDS (Index System).
○   These scanned and indexed bill pages were imported into System Z.
○   The sub-contract system package was dispatched to sub-contractors in three ways; email, hardcopy and in CD format.
○   After receiving the package back from the sub-contractors via post, the estimator evaluated them manually.

The whole tender process in producing the sub-contract document was repeatable for the new sub-contract documents which allowed organizations to repeat successful practices developed on earlier projects. The overall process can be described as well structured, because the planning and tracking of the tender process was done rigorously; earlier successes could be repeated and past failures avoided. For example, the information in the earlier projects could be used for future sub-contract documents, and improved information sharing, i.e. templates, sub-contractor selection, and evaluation, etc. Policies for managing sub-contract quotations and procedures to implement those policies were established. The procedure for managing sub-contract documents was improved as the work progressed.

The Criteria Relating to the Readiness Model in Level 3

*   Practice; Business processes were standardized organization-wide.

Target Status (the Readiness Level that Organizations Desire to be in the Future)

Practice

*   4th phase (future-prediction) – *100% electronic*
  ○   Main tender document to be received electronically
  ○   Estimator should select which pages are to be indexed with the CDS Index System and then imported into EDMS
  ○   Then the sub-contract package should be downloadable via the website.
  ○   Receive package back from sub-contractor electronically in the original format, and the estimator evaluates electronically.

E-tendering solutions will provide a single point of information across an organization. This advantage also extends beyond the organization while all sub-contractors obtain the same documents from a single source of information, and disputes caused by inconsistency of documents in the traditional tendering processes can be avoided. This will increase the degree of satisfaction of all parties involved, and maintain the level of harmony across the supply chain. The system can have an e-mail trigger process control to alert that someone has made changes to the original document. Thus, the company gains the capabilities to monitor and maintain the quality of the original document up to certain standards. All tender projects should use the same well-defined process, and therefore the quality, time, and cost could be measured, compared and standardized across tender projects. When the quality, time and cost do not match intended standards, action should be taken to correct the situation. In addition, e-tendering solutions should also facilitate the audit trail of tendering processes throughout the cycle, which should improve the transparency and fairness. The automatic tender reporting functionality should ensure the tenders are evaluated on the same platform.

The Criteria Relating to the Readiness Model in Level 4/5

- Practice; business process should be well defined and capable of setting quality and measuring the products/services.

The results of process analysis are as shown in Table 3.

PEOPLE

Skill

Organization C had an excellent track record in training and education of their employees. In 1992, Organization C was the first national contractor awarded the Investors in People Award (IIP), and was awarded a National Training Award in February 1996 for their in-house management development programme. In 2003, Organization C achieved Matrix Standard Accreditation for their internal training programmes. The training ranged from induction, IT skills, health & safety for its managerial staff and also on-site operatives and craftspeople. The Matrix Standard

Table 3. Process: Practices Analysis

Status	Prior to	At termination/Current	Target	Gap
Maturity Level	2	3	4/5	3,4,5

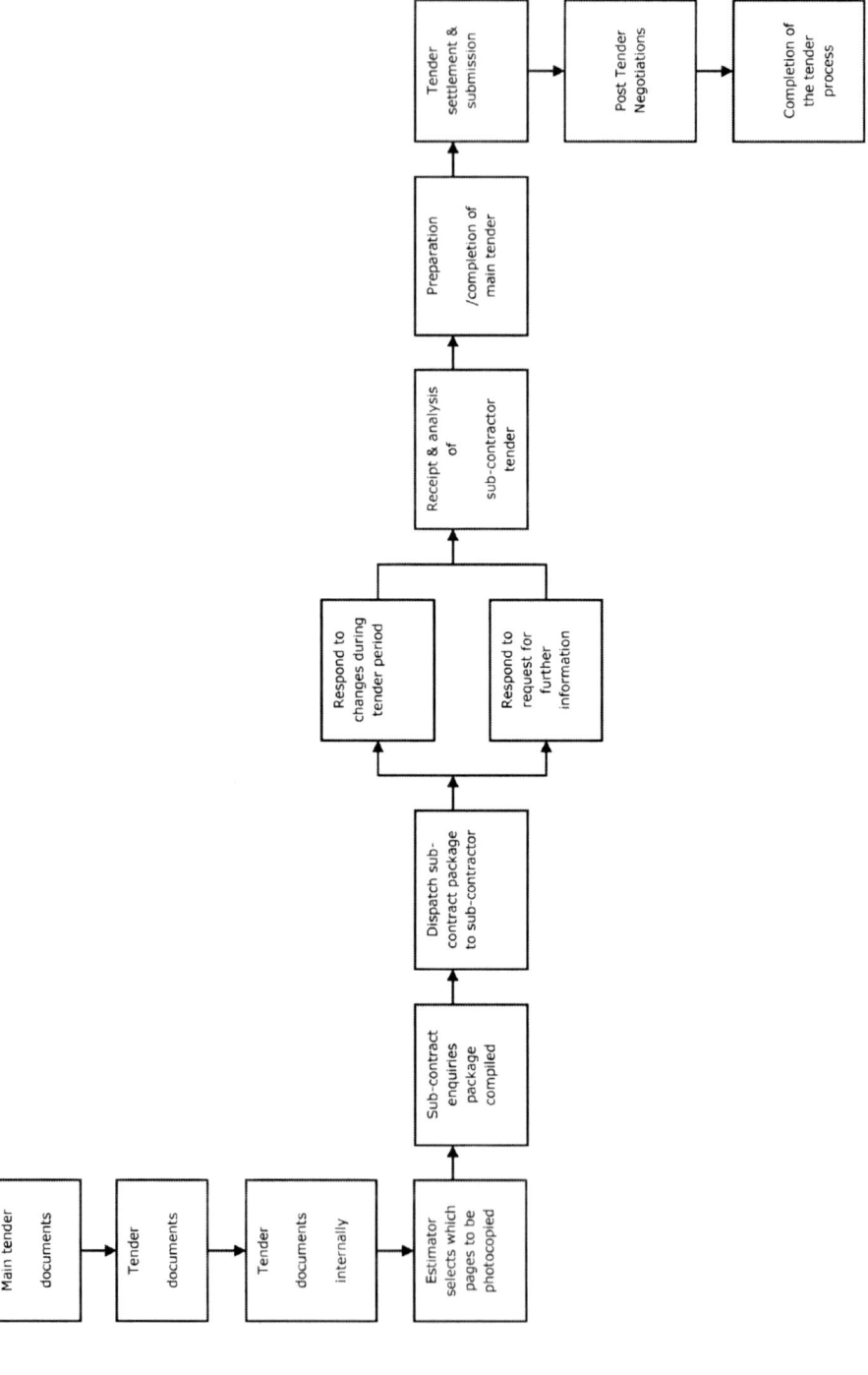

Figure 5. Status prior to project (Process A)

Figure 6. Status prior to project (Process B)

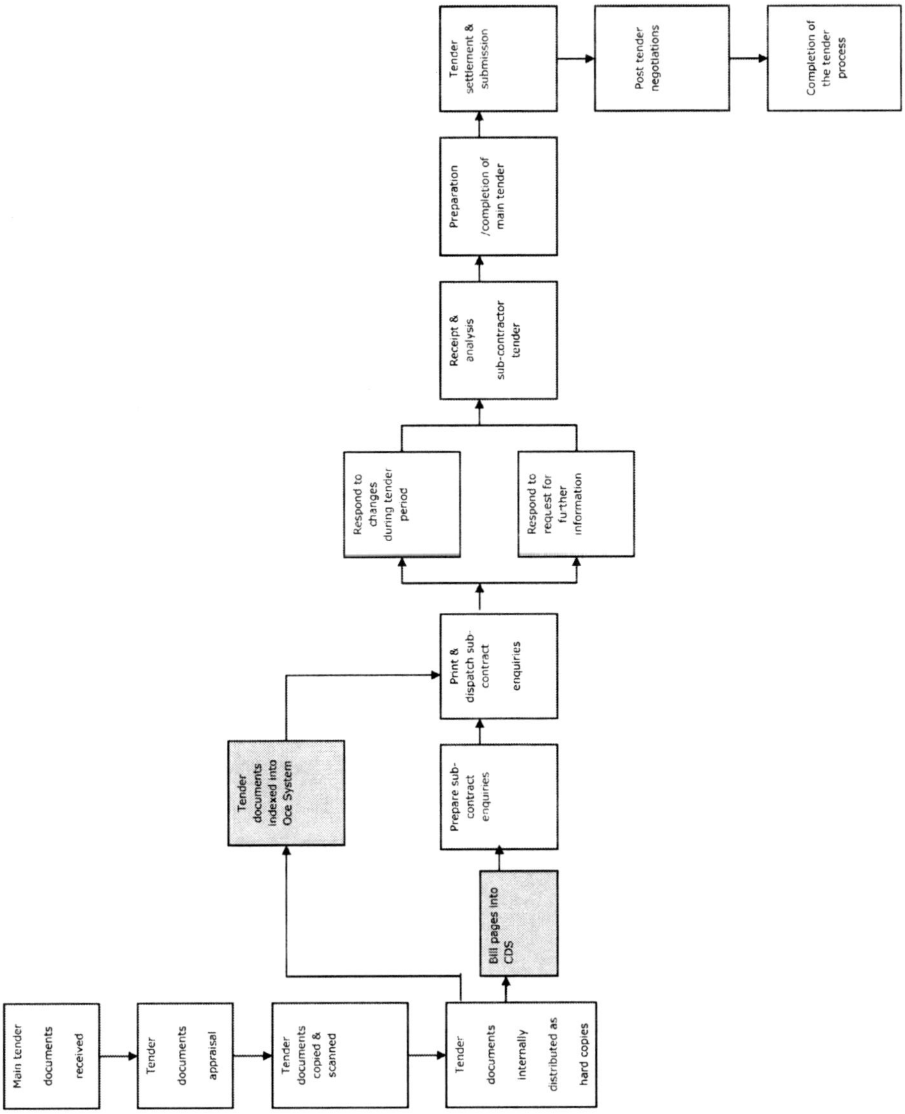

Figure 7. Current status of the project (Process)

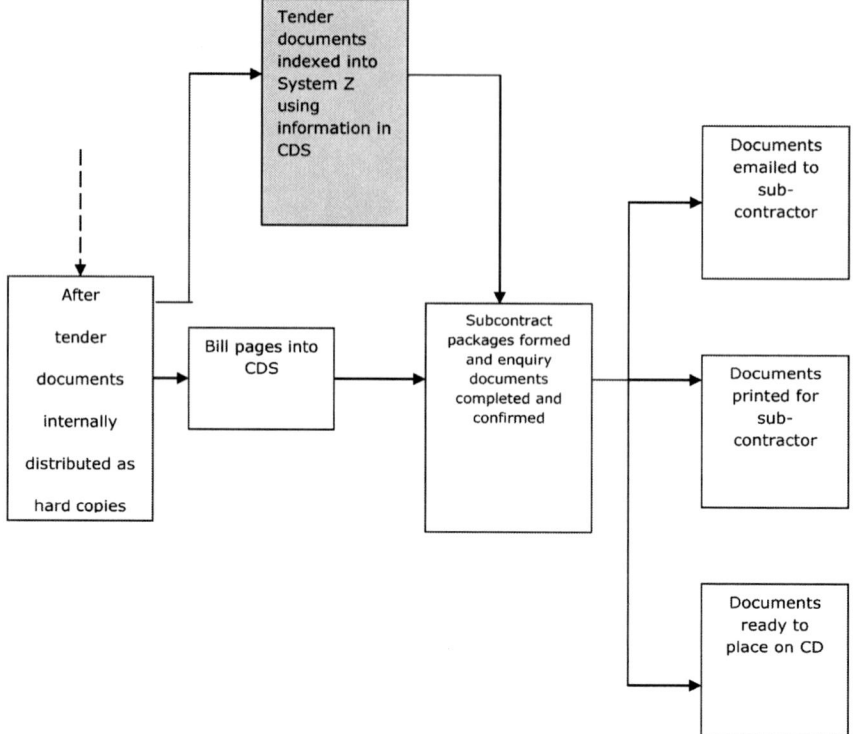

is the national quality standard for any organization that successfully manages information, advice and guidance on learning at work. In February 2007, the practice was again accredited for the Matrix Standard. According to organization's Learning Manager, The *Matrix* Standard helped the practice to recognize the successes of individual teams and encouraged consistency across the business in the application of policies and procedures. Organization C's employees voted it the "Best Place to Work in Construction" at the 2006 Contract Journal Awards. In 1994, Organization C managed to form a partnership with Organization A to set up a unique management development programme. The main elements in this partnership were work based experiences that were developed to improve the manager's performance with an incentive of a postgraduate diploma, and with an option to progress to a Master's degree in Construction Management. The key features of the programme were:

- Company selection of participants
- Recognition of prior course attendance and experience
- Work-based learning, supported by mentoring
- Core company-based modular course

Figure 8. Target Status (readiness level that organizations desire to be at in the future - Process)

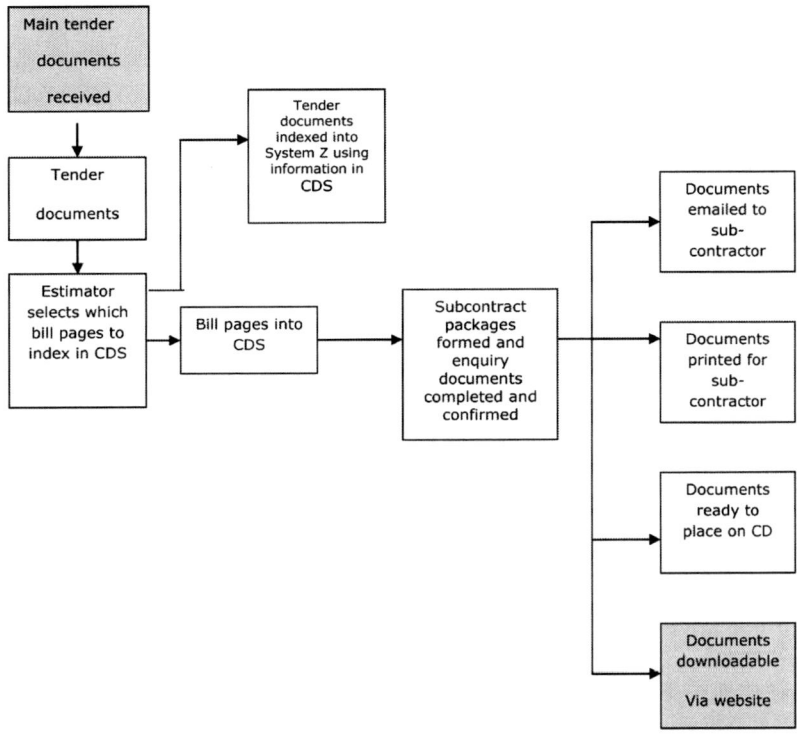

- Involvement of senior managers throughout the programme
- Achievement of academic standards and a recognized award
- Results measured in both business and academic terms

Status Prior to Project

Types of Skills

The in-house expertise had the capability to develop and maintain IT/IS such as programming, analysis, being able to install and manage off-the-shelf, readymade packages, etc. The regular users of IT/IS had the capability of operating a technical standalone package such as AutoCAD, CAD estimating, Agresso, etc. Technical skills were needed in order to operate such packages. For example, a designer had to gain technical knowledge in producing drawings and/or the surveyor must have the ability to interpret the drawings in order to produce bills of quantities. Apart from that, the normal packages used were word processing, spreadsheets, databases, etc. The skills needed to operate such packages were familiarity with basic Windows functionality. If there was any crisis or problem during operation of the application, the users were expected to seek guidance from in-house experts.

Capability Building

There were many training programmes conducted within the practice. However, the training was more focused on specific business units. For example, the training for using an estimating package was conducted for estimating department staff only. There was a culture of knowledge sharing that existed within the practice, but it was also limited to specific business units. On some occasions, there were centralized training programmes dedicated to all staff within the practice. Many staff benefited from the special training scheme existing between the practice and Organization A.

The Criteria Relating to the Readiness Model in Levels 3 / 4

- Skills; considerable technical and project management skills
- Capability building; centralized integrated training program within organization.

Current Status of the Project

Types of Skills

In-house experts have developed and gained project management experience during System Z's development. Vendor C plays a vital role in providing advice and transferring knowledge to in-house experts, and they expressed their confidence in developing and maintaining the IT/IS in the future. The interaction between departments was more frequent particularly in the sub-contract tendering process. For example, the selecting, checking and inserting of relevant documents such as drawings, preliminaries, and bills of quantities required more frequent interaction between departments. The IT skills needed to operate the system was the familiarity of basic Windows functionality, and understanding the tendering process, and the practice also develops the skills matrix for training and development of IT skills (Table 4).

Capability Building

There was an integrated plan of training that was developed throughout 2006/2007 for all staff, and the main aim of this training was for closer and correct IT integration with the supply chain. This method of training program included work based experience, workshops, one-to-one sessions, and seminars, which include:

Table 4. Organization C's IT skills matrix

ICT Department Plan 2006/2007		Introduce Corporate Document Management System	Standard Project Collaboration System adopted	CDS Conversion	All site staff to laptops	Extend use of Citrix to sites	Dashboard	3D/ Visualization	Handheld system for safety	Handheld for snagging	Servers on all sites over £m	Mobile Data Connections for sites
	Team Leader	ZT	PH	JO	NT	MB	SCJ	RS	SS	SS	SS	SS
	Team No	1	2	3	4	5	6	7	8	9	10	11
	Champion											
Closer IT Integration with supply chain			*					*				
Correct IT in the right place at the right time					*	*						*
Improved process through use of appropriate IT		*		*	*	*	*	*	*	*	*	
Measure		In use on multiple sites and in offices	In use on multiple sites	Progress measured against project plan	100% complete by end of financial year	Citrix in use on sites	Dashboard requirements specified	Requirements specified	System in use on first site	System in use on first site	Servers in use	Solution found for initial site setup period
Target date		Jun 07	Jun 07	Apr 07	Jun 07	Jun 06	Dec 06	Dec 06	Jan 07	July 06	Jun 07	Sep 06
Performance: Red, Amber, Green												
Team Members: Zoe Turnbull, John Often, Steve Slater, Steve Burton, Neil Tennant												

- Introducing document management systems – workshop/work based experience
- Standard project collaboration system – workshop/work based experience
- CDS conversion - workshop
- All site staff to laptops – one to one session
- Use of Citrix on sites - workshop
- Dashboard - workshop
- 3D/Visualization - workshop
- Handheld system for safety – one to one session
- Servers – one to one session
- Mobile data connections – workshop

The Criteria Relating to the Readiness Model in Level 5

- Type of skills; gain cross disciplinary experience to support supply chain activities
- Capability building; knowledge sharing and knowledge management.

Target Status (the Readiness Level that Organizations Desired to Achieve)

Types of Skills

In-house expertise was expected to improve the capability to develop future IT/IS systems. Integration of key skills should be embedded within the future e-tendering system, and each department should be aware of what the others are doing. All users are expected to be able to manipulate and maximize the potential of IT to the practice.

Capability Building

Establish collaboration and knowledge sharing within and outside the organization, and the practice also will establish a partnership within industry and academia.

The Criteria Relating to the Readiness Model in Level 6

- Type of skills; Acquiring skills to develop IT core capabilities
  Capability building; Based on inter-organizations sharing experience strategy.

The results of skills analysis are shown in Table 5.

72

Table 5. People: Skills Analysis

Status	Prior to	At termination/Current	Target	Gap
Maturity Level	3/4	5	6	4,5,6

Roles and Responsibility of IT Staff

The IT department was structured into two units:

- Software Development – consisted of 10 staff including a manager. Their roles were to determine requirements for the system, set up a steering team to run system development (led by the senior management champion, the user team leader and other members depending on the size of the project, and relevant people from software development), and write the specification. When the specification was agreed, it went through different levels of development. User teams were formed and they tested and wrote the user guide for the system. There was user acceptance before the system was implemented for a 3 month review.
- ICT Support – consisted of 14 staff including a manager. Their role was to investigate and conduct the pilot project, test, implement, improve facilities and IT performance to support business.

Both managers have to report to the ICT Director.

Status Prior to Project

Position of IT/IS Head

The Head of IT/IS position evolved from 1991 with the appointment of a software developer. Then in 1996, the company promoted the software developer as the head of the computing unit; a unit under the Finance Department. In 1999, the computing head was appointed as head at the new IT department.

Roles

To provide technical support and purchasing on software and hardware.

The Criteria Relating to the Readiness Model in Level 3

- Position of IT/IS Head; Technical oriented Manager
- Roles; purchasing and centralized IT/IS activities.

Current Status of the Project

Position of IT/IS Head

In 2003, the computing head was promoted to ICT Director, and the IT department was divided into two sections; ICT support and Software Development. The ICT Director reported to the Finance Director who was a member of the board of directors.

Roles

- The role of IT staff expanded
- To define ICT strategy in line with business strategy.

The Criteria Relating to the Readiness Model in Level 5/6

- Position of IT/IS Head; IT Manager at senior management status
- Roles; IT/IS strategy and business strategy.

Target Status (the Readiness Level that Organizations Desire to be at in the Future)

Position of IT/IS Head

The IT/IS Director was going to be promoted with full membership of the board of directors as Chief Information Officer (CIO) in the near future. The ICT support Manager will be promoted to ICT Director.

Role

- IT/IS strategy and business strategy.

The Criteria Relating to the Readiness Model in Level 6

- Position of IT/IS Head; IT Head with full member of board of director
- Roles; IT/IS strategy and business strategy.

The results of roles and responsibility of IT Staff analysis are shown in Table 6.

Table 6. People: Roles and Responsibility of IT Staff Analysis

Status	Prior to	At termination/Current	Target	Gap
Maturity Level	3	5/6	6	4,5,6

User Involvement

In most cases, the users were involved in high cost and large scale projects. Normally, the users got involved in groups, and not as individuals; and this group or committee was established on a temporary basis.

Status Prior to Project

User Involvement

The users contributed to the IT/IS project by expressing their needs and requirements towards a proposed system. Normally, group users within the practice would be consulted when the decision to purchase, or develop a system was made. However, in most cases a group of users were involved on an ad-hoc basis, and were only involved in high cost IT/IS projects.

The Criteria Relating to the Readiness Model in Level 3

• User involvement; focus group consultation on purchasing and centralized IT/IS activities.

Current Status of the Project

User Involvement

User representatives came from groups of people who were going to eventually be involved in using and maintaining the system. For example, for the snagging system, all operation staff were involved (such as the construction manager and the site manager). Also, for System Z, all the estimators were involved in prioritizing the system's requirements. However, some user representatives from other departments were also selected randomly to participate – user involvement was limited to purchasing that system and was not related to the IT/IS strategy or business strategy.

The Criteria Related to the Readiness Model is in Level 3

User involvement; focus group consultation on purchasing and centralized IT/IS activities.

Target Status (the Readiness Level that Organizations Desire to be at in the Future)

User Involvement

The group of users was expected to be formed into an e-tendering system. This was because user representatives came from various departments involved in the e-tendering system, such as users from the estimating department, finance department, accounts department, etc.

The Criteria Relating to the Readiness Model in Level 4

*   User involvement; focus group consultation (managing information across company)

The results of user involvement analysis are shown in Table 7.

WORKING ENVIRONMENT

Organization Behaviour

Users involved in the system functional and non-functional requirements definition for System Z were solely from the estimating department. Consequently, the users from the estimating department were more comfortable with the new system, while the other users seemed reluctant to change their old practices at first; and the trial time of three months prior to System Z's actual implementation could have been the factor that reduced resistance to change. Generally, there were two patterns of acceptance and resistance towards the new IT/IS implementation within the practice:

Table 7. People: User Involvement Analysis

Status	Prior to	At termination/Current	Target	Gap
Maturity Level	3	3	4	4

- Replacing the manual to computer system era, when the practice wished to change the old paper-based practice to computers. The resistance to change occurred across almost the entire practice. The only staff who accepted the changes at first were the technical staff.
- Replacing computer system to computer system era, when the practices wished to change the half paper based practice to the new improved IT/IS. The resistance to change occurred among senior staff rather than junior staff.

Training sessions after the implementation play a vital role to give confidence and understanding among the users.

Status Prior to Project

Characteristics

There was a lack of co-operation between the users and in-house experts and the IT/IS development was more focused on technical perspectives such as time and quality, rather than human perspectives, which were only considered after the system was already in place. For example, there was no user consultation regarding what kind of needs and requirements they wanted for the new IT/IS. The needs and requirements of the users were only identified during the training session after the system became live, creating resistance among the users.

The Criteria Relating to the Readiness Model in Level 2

Characteristics; technology led, informatics culture.

Current Status of the Project

Characteristics

The practice learned from what they experienced previously regarding user consideration during IT/IS development. The practice has made lots of preparations to introduce System Z to the users, including planning training and awareness sessions and beginning to seek the user's needs from the estimating department to state what they want in System Z. Random users from other departments also participated. The practice focus is to improve the sub-contracting tendering process and an awareness session conducted prior to System Z's actual implementation helped the practice to reduce resistance to change. The users seem prepared and had an idea what System Z would look like. The transition period between the old sub-contract tendering process to the new one went very smoothly, without problems.

The Criteria Relating to the Readiness Model in Level 4

Characteristics; the focus is on ensuring that all organization functions and operations are running smoothly.

Target Status (the Readiness Level that Organizations Desire to be at in the Future)

Characteristics

It is expected that the practice might get the involvement of sub-contractors in order to enhance System Z to a full e-tendering system. For example, the sub-contractor can express what kind of function they want to be inserted in the e-tendering system. The marketing aspect was the main factor to ensure the e-tendering system ran successfully and ensured better services were offered to the clients. The e-tendering process will improve the image of the practice.

The Criteria Relating to the Readiness Model in Level 5

Characteristics; IT/IS used to provide better products/services to the customer
The results of organization behaviour analysis are shown in Table 8.

IT Policy

The practice seemed to maintain the policy of centralized IT/IS activities throughout the organization. Prior to the IT department being established, the software developer was managing all aspects of IT/IS activities. The practice also gave some space for their business units to make a recommendation on what system they wanted. However, the IT department did all the purchasing and development of the system.

Status Prior to Project

IT/IS Activities Control

Prior to the IT department being established, the software developer controlled all aspects of IT/IS activities, before the IT department took over the role.

Table 8. Work Environment: Organization Behaviour Analysis

Status	Prior to	At termination/Current	Target	Gap
Maturity Level	2	4	5	3,4,5

The Criteria Relating to the Readiness Model in Level 3

Centralized IT/IS infrastructure and application policy.

Current Status of the Project

IT/IS Activities Control

The IT/IS department fully control IT/IS activities, including infrastructure and applications.

The Criteria Relating to the Readiness Model in Level 3

Centralized IT/IS infrastructure and application policy.

Target Status (the Readiness Level that Organizations Desire to be at in the Future)

IT/IS Activities Control

It is expected that the practice will maintain the centralized IT/IS activities policy.

The Criteria Relating to the Readiness Model in Level 3

Centralized IT/IS infrastructure and application policy
   The results of IT policy analysis are shown in Table 9.

Leadership

The Managing Director is knowledgeable in IT, and senior management consider IT/IS as the vital element that leads to competitive advantage. The idea of developing the new IT/IS to improve the sub-contract tendering process came from the senior management. Senior management actively participated in meetings and discussions between the users, the in house expertise and the vendor during System Z development, and was very concerned about the impact of System Z on the number of

Table 9. Work Environment: IT Policy Analysis

Status	Prior to	At termination/Current	Target	Gap
Maturity Level	3	3	3	Improvement

contracts awarded to the practice. The Managing Director and ICT Director were among the speakers during the System Z awareness programme - communication about System Z was mostly done through email, newsletter and memo. In addition, the practice also developed an intranet system as the practice's publishing mechanism to publish the latest information and also create feedback forms for any issue.

Status Prior to Project

Communication

No standard way or communication plan was being applied within the practice, as everything was done on an ad-hoc basis.

Participation

The senior management was in the position to initiate the IT/IS project and prioritize which IT/IS developments should be in place at first. However, their participation was on an ad-hoc basis.

The Criteria Relating to the Readiness Model in Level 2

- Communication; no established standards in place for communication planning.
- Participation; ad-hoc participation basis.

Current Status of the Project

Communication

No standard way or communication plan was being applied within the practice, as everything was done on an ad-hoc basis.

Participation

The senior management was very concerned with the effectiveness of System Z to the practice on being able to win contracts; and measured the effectiveness of System Z by the number of contracts awarded.

The Criteria Relating to the Readiness Model in Level 2/5

- Communication; no established standards in place for communication planning.

- Participation; Participate in performance measurement of IT/IS project.

Target Status (the Readiness Level that Organizations Desire to be at in the Future)

Communication

The practice is on the way to developing a communication plan for managing construction site management across the organization. This is expected to influence the communication of IT/IS activities in the practice.

Participation

The senior management participation on the IT/IS project is expected to continuously improve.

The Criteria Relating to the Readiness Model in Level 3/6

- Communication; Organization-wide policy and standards for communication.
- Participation; Participate actively in continuous improvement of the IT project.

The results of leadership analysis are shown in Table 10.

SUMMARY AND FINDINGS

Almost all the attributes identified were not at lower level 1. The practice achieved maturity in three areas – skills and leadership are identified at level 5, and roles & responsibility were identified at the top of the level. There was no improvement found in the areas of IT policy, where the practice was identified at level 3 at three different situations; prior to project, current status of the project, and target status (the readiness level that organizations desire to be at in the future). This indicates

Table 10. Work Environment: Leadership Analysis

Status	Prior to	At termination/Current	Target	Gap
Maturity Level	2	2/5	3/6	3,5,6

that the practice is not ready or does not have any plan to give the authority to the business unit to make decisions on the purchasing/development of IT/IS. Two attributes did not improve as part of this new system implementation; user involvement was found to be at the same level (3) prior to the project and at termination/current status, and leadership was also found to be at the same level (2) both before and after the project.

The practice was managing well with their sub-contract tendering process maturity improvement, which follows the expected sequence according to the readiness model; prior to project - *manually* (level 2 - process), current status of the project-*half manually* (level 3 - process) and target status (readiness level that organizations desire to be at in the future) - *fully electronic* (level 4, 5 - process).

Figure 9. Summary of the assessment of System Z's implementation

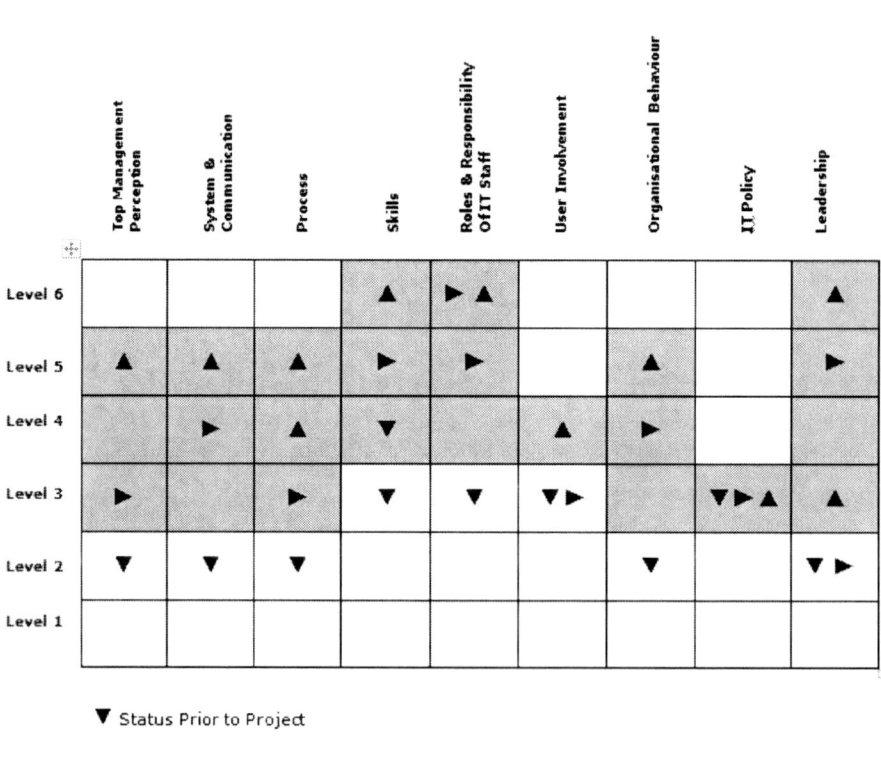

▼ Status Prior to Project

► Status at Termination/Current

▲ Target status (the readiness level that organisations desire to be at in the future)

▨ Gap

Prior to the project, the in-house IT expertise had the capability to develop and maintain IT/IS, such as programming, analysis, being able to install and manage off-the-shelf, readymade packages, etc. The regular users of IT/IS had the capability to operate technical standalone packages such as AutoCAD, CAD estimating, Agresso etc. There were many centralized training programmes conducted within the practice (level 3/4 - skills). The current status of the project is that the interaction between departments is more frequent, particularly in the sub-contract tendering process. The practice also developed the skills matrix for the training of IT skills to develop IT core capabilities in the future IT/IS (level 5 & 6 -skills).

The practice acquired an adequate capacity for the IT department. Prior to the project, the IT Manager was appointed to Head of IT Department, with the role of purchasing and centralizing IT/IS activities (level 3 – roles & responsibility of IT staff). The IT department expanded gradually and was split into two units; software development and ICT support. The IT Manager was also promoted to senior management status to facilitate the contribution of the IT department towards the IT/IS strategy and business strategy (level 5, 6 – roles & responsibility).

It was identified that there was no improvement in the maturity level for the users involved in the IT/IS project, prior to and at the termination/current situation, where the users identified got involved in a focus group consultation on purchasing and centralized IT/IS activities (level 3 – user involvement). There was no improvement in IT/IS activities control, since all IT infrastructure and applications remained centralized (level 3 – IT policy).

The summary of the assessment of System Z's implementation according to the readiness model can be found in Figure 9.

CASE STUDY ADOPTED FROM

Salleh, H. (2007), "Measuring Organisational Readiness Prior to IT/IS investments", PhD Thesis, School of the Built Environment, The University of Salford, UK.

Chapter 4
Building for the Future:
Systems Implementation in a Construction Organization

Hafez Salleh
University of Malaya, Malaysia

Eric Lou
University of Salford, UK

EXECUTIVE SUMMARY

This chapter provides the IT readiness assessment for before and after scenarios of IT systems implementation in a construction consultancy company providing multi-disciplinary services for the construction industry throughout the United Kingdom. The services offered include building surveying, quantity surveying, project management, civil and structural engineering design, and mechanical and electrical engineering design, among others. On application of the maturity model it was found that the overall processes for managing information are improving since the introduction of the new IT system. Prior to the project, the development of IT/IS was driven to perform daily work tasks that required the company to run a business. The new systems has streamlined the organization-wide communication, which the previous system did not have the capability to do, and to reduce cost for document reproduction. The level of IT skills prior to the project was relatively low; the introduction of the new system has helped the company to increase their staff's IT skills.

DOI: 10.4018/978-1-61350-311-9.ch004

BACKGROUND AND HISTORY

Organization B is a construction consultancy company providing multi-disciplinary services for the construction industry throughout the United Kingdom. The services offered include building surveying, quantity surveying, project management, civil and structural engineering design and mechanical and electrical engineering design, among others. Organization BBBBB was established in 1941 and operates from their offices in four different cities. The organization turned into a Limited Liability Partnership (LLP) in April 2006, with an annual estimated turnover of £12 million. Organization BBBBB employs 220 staff across 4 offices, of which 120 staff are located at the Head Quarters of the company. No specific department exists in their organizational structure. Instead, the organization operates in groups, but not strictly by discipline. For example, one group consists of multiple disciplines, and anyone can be a group leader. The disciplines are as follows: Building Surveyor, Quantity Surveyor, Project Manager, Employers Agent, Architect, CAD, Mechanical and Electrical Engineers, Civil and Structural Engineers, Planning Supervisors. There are three layers of management within Organization BBBBB's organization structure; The Executive Group, The Senior Management Group and The Practice Group. Organization BBBBB's organization structure is shown in Figure 1.

Sequence of Events

In November 2004, the management of the company discussed the need to replace their existing Database Management System (DMS), which had reached the limit

Figure 1. Organization BBBBB's organizational structure

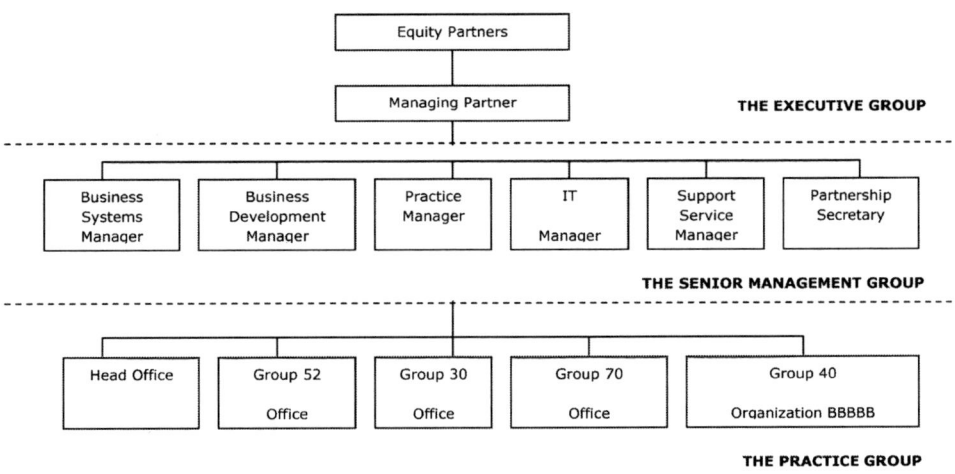

of its development potential. In doing this, they had to recognize the need for a coherent approach to their IT, and the need to bring all systems under one control.

In 2004, the management wanted to have access to all parts of the knowledge system through one front end. Further discussions established that they wanted to include a Knowledge Management System (KMS), and added Human Resource (HR) functionality to the proposed system. The organizational business system managers suggested what the practice really needed was a single piece of software that had the capability to manage all the functions within the organization. The management agreed, and asked for a time frame. The project team was then established and comprised of a business system manager as head, and IT manager, a database programmer and a business development manager; and started to evaluate the products available in the market. During the product evaluation, the project team identified products which could do more than just DMS functions, and in fact provide them with a Client Relation Manager (CRM) which is essentially an advanced address database with functions that make the information both accurate and usable, and also an extranet function which makes all documents retrievable from remote locations and an intranet for internal information distribution. The whole function combined to provide a sophisticated Knowledge Management System (KMS). This is what they thought many people had been requesting for some time. The project team identified the preferred vendors, and five vendors were invited to present their products. One of the products that impressed the project team was the product from Vendor B, and this vendor was invited to present one of their products called Workspace ™. The presentation was attended by the managing partner, business system manager, IT manager and the database programmer. In addition to Workspace ™, the project team also reviewed other products by attending seminars, conferences, and seeking views from the companies who had implemented similar systems.

The product evaluation process took up to a year (then 2005). Finally, the project team decided to select the product from Vendor B as it met the specification of the required system for the company to bring all systems under one umbrella. A payment was made to Vendor B of about £200,000 and internal development costs were going to be added to this later. After signing the contract with the vendor, the business systems manager came out with a methodology (programme) of software development, including the necessary adjustments to the software and the time frame. Both these were represented in terms of business processes.

- **Initial phase**
  - looking at the corporate appearance of software
  - In house staff designed this corporate appearance
  - Looking at the extra functions to be added to the software

- ○ Within three months after that the management decided there was an urgent need to be looking at a human resources (HR) package
- **Second phase**
  - ○ Started to evaluate a few human resource (HR) products to include human HR functionality to the workspace.
  - ○ Started to load data into the HR product
  - ○ Human resource managers complained the software was not functioning as expected.
  - ○ The vendor who was supplying the HR product was very unhelpful, and imposed extra charges on every extra hour they spent on it which totally contrasted with Vendor B.
  - ○ Vendor B put in a lot of time to put the system in order without extra charges
- **Third phase**
  - ○ Terminated the contract with the vendor that supplied the HR product
  - ○ Asked Vendor B to incorporate the HR function to their product
  - ○ Started the development process
  - ○ Steering committee (SC) formed to represent all functions and levels in the practice
  - ○ Gave a presentation to SC about what the team had been doing with Vendor B so far, and sought their comments and agreement.
  - ○ The SC were asked to look at and test the system, but it did not go well
  - ○ There are several stages of testing
    - ▪ The in-house staff renamed the product as 'System Y' (for corporate appearance)
    - ▪ Selected staff tested the system – open, enter and save documents
    - ▪ It started in the Winchester office
    - ▪ Management expressed concern over time delays and fears of the system crashing if large numbers of users used it simultaneously
    - ▪ The business development group also participated in testing the system and commented on it.
    - ▪ Training managers went to each office branch and provided one to one training to each member of staff
    - ▪ All the members were trained to create and save documents in System Y. This is a basic level of training, and staff were certified to use the basic operations of System Y.
    - ▪ Some of the partners said they wanted to work remotely, for example at home. The business system manager suggested it would be possible if a scan function could be added to the system in order to review documents off site.

- **Fourth phase**
  - This phase focused on the issue of how to maximize the capabilities of the system, particularly involving people issues
  - All suggestions and feedback were reviewed
  - A business case and future priorities were produced
  - Final specifications were then produced

2006

- System Y went 'live' in February 2006 with basic functions.
- System Y was a document management system (DMS) designed to facilitate the management and administration of corporate documents within Organization BBBBB.
- There were 4 main entities/zones that were used within System Y (see Figure 3):
  - Documentation – all the documents produced were stored.
  - Contact – all contact information held on both internal staff and external clients.
  - Project – all information to projects was stored.
  - Organization – all company information was stored.

2007

Figure 2 illustrates the timeline for System Y's development.

Figure 2. System Y's development timeline

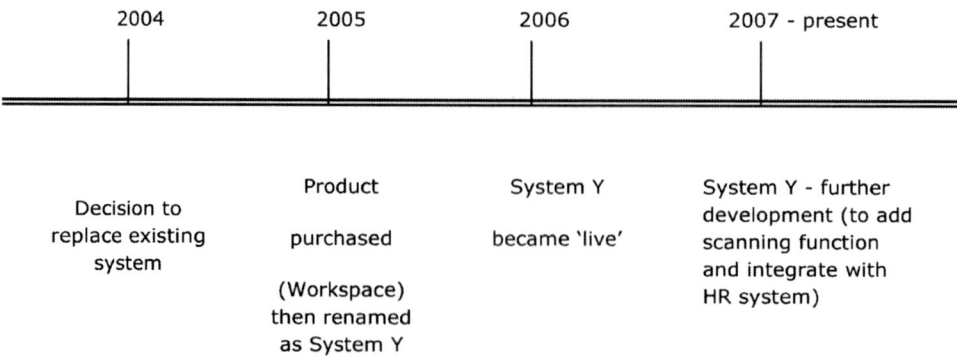

Figure 3. System Y

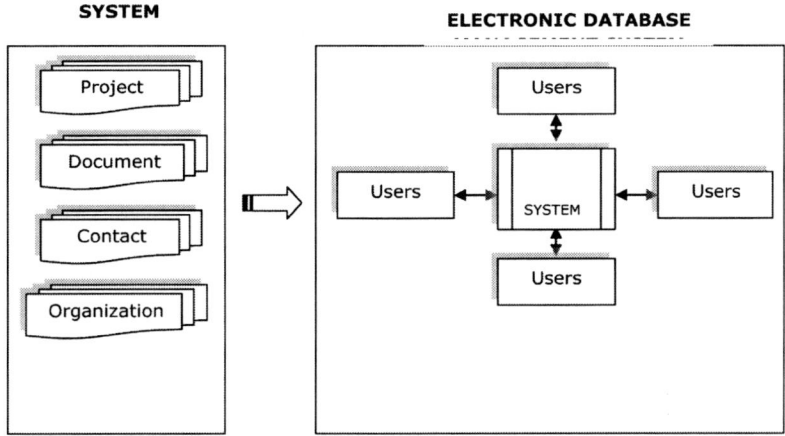

ANALYSIS AND DISCUSSION

The following section addresses each attribute of the four domains of the readiness model within the context of Organization BBBBB.

IT INFRASTRUCTURE

Top Management Perceptions

The management of the company which includes the executive group and senior management group established the vision of strategic thinking towards the general current and future business plan, and the level of resources needed, including staff, training, IT system development, equipment, and identifying opportunities and risks to the business, etc. In addition, any project including construction development and IT system development was planned in terms of its own individual requirements. In terms of IT system development, the business systems manager was responsible for ensuring that IT systems were suitable with business needs and were properly administered and maintained. Once Organization BBBBB became a Limited Liability Partnership (LLP) in April 2006, it facilitated more staff participation in decision making in the company's strategic planning. The *status prior to project* refers to the period prior to 2006 in which the existing Database Management System (DMS) was still in operation. The *current status of the project* refers to when System X became 'live' in 2006 and up to the present day. *Target status* refers to the readiness level that organizations desire to be at in the future.

Status Prior to Project

Drivers

The development of IT/IS is purely to support task requirements and to improve the internal efficiency of various workgroups such as designing, costing, procurement, programming and reporting that is required for the company to be in business. The other IT systems (word processing, spreadsheets, CAD and visualization, computer aided cost estimating and planning and scheduling) are solely independent, off-the-shelf, and primarily related to support individual work tasks. The decision to purchase any system/application was according to the necessity to perform single work tasks. No feasibility study towards the direction of IT/IS development within the company had been done. The need to purchase the system/application was also triggered by looking at systems used elsewhere. A very small IT/IS plan was developed, however its nature was on an ad-hoc basis.

Systems Requirements Definition

The product vendors played an important role in influencing the company to purchase the system/application. The company heavily depended on the vendor to set the functional requirements of the system purchased, while the management determined its non-functional requirements in terms of speed, quality and cost. The applications/systems used within the company were solely dependent on standard featured software and hardware vendors, and operated on a DOS platform. The company did not have any possibility to make alterations or enhancements to the original product they used. The planning, designing and installing of the products was done mostly by vendors.

The Criteria that Relates to the Readiness Model in Level 1

*   Driver; copying/work task requirements
*   Systems requirements definition; heavily depends on vendor.

Current Status of the Project

Driver

There is an issue relating to the consistency of information storage and retrieval organization-wide. Everyone has their own separate directory and lack of information sharing is obviously emerging. Information queries from the London office take time for a response. The other issue was that all the documents were saved in individual standalone PC, and therefore it took time to get the information needed, and the possibility of lost documents was relatively high. To solve this problem, the management decided to develop a centralized database that was accessible within the company. In doing this, the company recognized the need for a coherent approach to their IT, and the need to bring all the systems under one central control. There was also a space issue regarding the allocation of the filing cabinet for storing the file documents. Cost of reproduction for project documents was also high. The company then decided to develop systems that have access to all parts of the knowledge system through one front end, and be managed by a central database administrator located in Orpington. For example, the applications used Quatopro, Auto CAD, Word, Excel, Visio, and Microsoft Project, and these would be linked together which could then be opened in a single application, 'System Y'. With this feature, the sharing of information and streamlining of communication and inter workgroups co-ordination became more efficient. The main reason to develop System Y was to streamline communication organization-wide, which previous systems did not have the capability to do, and so reduce the costs for document reproduction.

System Requirements Definition

A lot of customization needed to be done to 'Workspace' ™. Organization BBBBB produced a methodology of software development including the necessary adjustment and customization to the software and time frame including renaming the product 'System Y'. The major customization of Workspace ™ was to fix the scanning function which facilitates document scanning and viewing. The customization work was done by Vendor B. All the functional requirements and non-functional requirements of System Y were compiled by in-house experts with vendor assistance, and the role of Business System Manager in System Y's development was interpreting the needs of the practice into the technological solution. Vendor B had a project manager dedicated to this practice and information flowed freely between them; there was also a helpdesk function built in. The development of IT/IS started to increase in a planned manner with management support.

The Criteria that Relates to the Readiness Model in Level 3

- Drivers; cost reduction, organization communication
- System requirements definition; mainly in-house with intervention from the vendor.

Target Status (the Readiness Level that Organizations Desire to be at in the Future)

Driver

There is great demand from the clients and project team members to monitor the progress of construction projects more closely and transparently. The future situation was to meet this demand by extending System Y's accessibility to the clients and project team members, and the clients and project team members were given access to System Y to track down the project progress and status. There was also demand from the clients to integrate System Y with their system. Along with these improvements, the agenda to increase internal efficiency was still moving by integrating System Y with the Staff Management System (Human Resource Management System). The future improvement was to add value to System Y by extending its accessibility along the company information supply chain project – this function will increase the organization's competitive advantage.

System Requirements Definition

The company was planning to expand their IT Department by employing more technical staff, particularly the systems analyst, business analyst, and programmer. In the future, the company intended to get full involvement of internal expertise throughout the system's development. The design and set-up of the IT infrastructure such as the operating system, networking, security and telecommunications, was the responsibility of the vendor.

The Criteria Relating to the Readiness Model in Level 5

- Driver; information supply chain project
- System requirements definition; Full in-house with intervention from vendor

The results of top management perception analysis are shown in Table 1.

Table 1. IT Infrastructure: Top Management Perception Analysis

Status	Prior to	At termination/Current	Target	Gap
Maturity Level	1	3	5	2,3,4,5

Systems and Communication

Before 1996, the company had a basic network system but only secretaries had access to it. Quantity Surveyors used stand alone PC's for measurement, etc. There was no internet access or email. Telephones were basic switchboard type and dictaphones were tape type. Photocopiers and faxes were also basic. In 2000, all computers were fully networked and the practice had 80% of staff using computers. Currently, the company has 100% computer access. The telephone, fax and scanner systems are all integrated and managed centrally.

Status Prior to Project

Focus

IT/IS development is to support processes that are required for the company to be in business.

These business operations are primarily related to accounting, designing, procuring, documenting, etc. IT is not critical to any aspect of the company business strategy. The development of IT/IS is purely basic and to support operational tasks and performing similar activities such as typing documentation, payroll, accounts, and human resources. The applications that performed single activities are solely independent, off-the-shelf, and primarily related to the core business such as CAD and visualization, computer aided cost estimating, planning and scheduling.

Network Communication

All the PC's are connected, limited within the group, all employees used their own/shared IT services such as printing and scanning, and all had their own directory to save documents in, that could not be shared by another. However, the documents were exchanged on a manual basis.

The Criteria Relating to the Readiness Model in Level 2

- Focus; co-ordinate project and business information
- Network communication; networks within workgroups in business unit

Current Status of the Project

Focus

IT development is devoted to centralized information activities, company wide. Current goals for IT focus are on cost reduction, improving quality and speed, and enhancing overall company effectiveness. The main aim is to store a document into a single database that is accessible by all staff within the company.

Network Communication

System Y provides a collaborative network for project management through a single login. It authorizes people at all branches/remote locations and the design team involved in the project to access all the information they need, regardless of where the information is actually located. Every PC is connected to each other and shares some common IT services such as printers, scanners and servers, and now the exchange of documents is mostly electronic.

The Criteria Relating to the Readiness Model in Level 3 / 4

- Focus; fully integrated software applications organization-wide
- Network communication; networks are used organization-wide.

Target Status (the Readiness Level that Organizations Desire to be at in the Future)

Focus

IT developments that extend services to support managing the supply chain to the customers and the design team. The IT/IS goals for IT focus on extending the market and the geographic reach. The focus is to make everyone use the same documents such as email, scanning documents, word application, spreadsheets, and drawings to fit around highly organized business processes.

Network Communication

The accessibility to the company database in System Y is extended to the clients and design team.

The Criteria Relating to the Readiness Model in Level 5

- Focus; manage information supply-chain

- Network communication; networks are used to improve supply-chain performance.

The results of system and communication analysis are shown in Table 2.

Table 2. IT Infrastructure: System and Communication Analysis

Status	Prior to	At termination/Current	Target	Gap
Maturity Level	2	¾	5	3,4,5

PROCESS

Practices

The overall processes within the organization were not defined and documented prior to the project. Also, prior to the project, everyone followed the same process (scope identified) in managing information; however the processes were unable to ensure quality and standards. Currently the information management has been standardized and everyone follows the same processes, which establishes quality and consistency organization wide.

Status Prior to Project

Practices

The process involved document processing over the whole lifecycle, consisting of incoming documents and 'in house' documents being held in hard copy and kept in dedicated file folders on the filing cabinet with a unique indexing system, documents being scanned when the job finishes, and held on optical disk. When the document was requested, staff could simply search in the filing cabinet according to the unique indexing system and make another copy of it using the photocopier and then the requested document was distributed to the particular staff member. Where documents or data existed on the computer then it could be accessed by anyone with the software on their computer, but searching for it was cumbersome and often required an email to staff members to find out where it was located. The process was repeated every time the same document was requested. There are several issues, which can be highlighted from this process:

- No document to record the document life processing.
- No assigned responsibility and authority for performing the tasks.
- Policies and procedures for managing project information throughout the company were identified and improved based on experience with similar projects.

- The organization typically did not provide a stable environment for managing project information, for example it lacked sound management practices and had ineffective planning.
- The process was unpredictable and constantly changed particularly when dealing with time and cost. The information process capability of the organization was unpredictable, because the management of project information was constantly changed or modified as the work progressed.

The Criteria Relating to the Readiness Model in Level 2

- Practices; business processes scope identified and improved as the work progresses.

Current Status of the Project

Practices

System Y can hold electronic copies of all documents, either in which they were created or in a scanned/pdf form and allows users to create, edit and manipulate documents with the same process and single resource as follows:

- The Find Zones within system Y enable the users to search for and manipulate data. The users are also able to create new projects, edit existing projects, create and edit documents, as well as create and edit contact information.
- The My Zone within System Y gives users access to the latest news, information, and correspondence, active files that users are currently working on, live projects, favorite links, and a list of all documents published, authored, or sent to the current users.
- There are a number of different ways the users can create new documents in System Y; firstly, a new document can be made with the listing templates according to their category, i.e. letters, reports, memos, etc. Secondly new documents can be created via the Project links. Thirdly, by creating a document based on an existing document within System Y; and fourthly, by using an existing document outside System Y as a basis for a new document by dragging a file from Windows Explorer into the Working Files area of System Y.

  The process can be summarized as follows:

- The standard process for managing projects information throughout the company are documented and standardized.

- System Y is created with the intention for everyone in the company to be able to store their work and record their activities in a controlled and standard way. For example, every document created such as emails, contact details, word applications, drawings and spreadsheets saved into a single repository can provide a simple and searchable interface to every document. This includes a full content index of most document types. That means the users can search everyone's work to see if something similar has been done before. Also, System Y keeps track of all information relating to the distribution process, thus, it is possible to find out when, how and why documents and drawings have been sent out. Apart from that, global views of all staff and their percentage utilization are also provided to allow the management to monitor the utilization for the entire company. Login and logout functionality exists in System Y, ensuring that two people cannot edit the same document at the same time.

The criteria relating to the readiness model in Level 3

- Practices; business processes documented and standardized

Target Status (Readiness Level that Organizations Desire to be at in the Future)

Practices

The future of System Y's development is to extend its accessibility to the external parties such as clients and project team members to get access into the system. With this function, the clients and project team can track down the progress of the related project, such as variation orders (changes of design, cost increase, extensions of time), payments, purchase orders, valuations, letters, minutes of meetings, etc. This would increase the client and project team member satisfaction and transparency of service being provided by the company, and would increase the level of harmony across the supply chain. Due to its transparency function, a project team is able to make complaints and enquiries in a more efficient and effective manner. A number of customer complaints and enquiries regarding the project progress might be set as a quality measure. However, the external parties' accessibility is limited to the 'project function' area only, which means they cannot view the other functions such as organizational documentation. The future of System Y's development is also intended to integrate with human resource system management for the efficient and effective management of human capital. The human resource management system functions as a single source of information, giving the employees' details such as personal histories, educational background, skills, capabilities, personal experiences, project handling experience, workload to payroll records. This therefore reduces the manual workload of administration activities, such as employee time and attendance, employee tax reports, pension plan to profit sharing and the capability for training and development, skills and capabilities management.

The Criteria Relating to the Readiness Model in Level 5

- Practices; business processes capable of setting quality and measuring the products/services

The results of business process analysis are shown in Table 3.

PEOPLE

Skills

Most staff gained basic IT skills; however, there are differences between junior and senior staff. The junior staff often request more IT facilities within the company, whereas the senior staff have the opposite view. The company is starting to employ young, dedicated people to fill the IT related positions. Network communication skills are becoming more essential due to great demand from staff and clients to invest in advanced telecommunications.

Status Prior to Project

Type of Skills

Most applications and systems within the company were off-the-shelf and developed by an external consultant/vendor. The in-house IT staff did not have the expertise to develop systems, and in-house staff did the installation of off-the-shelf and low level maintenance. The complex IT maintenance was done by an external consultant/vendor. Most application systems use basic IT functions such as switch on/off

Table 3. Process: Business Practices Analysis

Status	Prior to	At termination/Current	Target	Gap
Maturity Level	2	3	5	3,4,5

computer, creating, saving, and printing documents. The application use is simple and basic and does not require much IT skill. Daily operation of applications such as word perfect and spreadsheets for creating letters, documents, correspondence, specification, project planning, estimating, bills of quantities, etc. Design drawings were done manually. Obviously, no networking and electronic communication skills were required since all the information was distributed through hard copy. The users and even technical staff required assistance from an external consultant in resolving unexpected computer problems.

Capability Building

There was no formal IT training conducted to improve staff IT skills. Users took their own initiative to improve their own IT skills such as learning among themselves and learning through job experience. Some even took external courses on their own. The company only sent technical staff for external training to improve their programming and analytical skills.

The Criteria Relating to the Readiness Model in Level 1

* Type of skills; basic skills
* Capability building; individual effort.

Current Status of the Project

Type of Skills

The original system Workspace ™ (renamed System Y) has multiple functions that were defined by Vendor B. The customization tasks of System Y, such as defining functional and non-functional requirements, were mostly done by in-house expertise along with Vendor B's expertise. In-house expertise was gained with the system development experience during System Y's development. It has to be appreciated that System Y operates for all level of users including senior management, middle management, administration, and site management for Organization BBBBB. Most users, particularly the surveyors, utilize and benefit from the functions within System Y that support many tasks for project management, such as:

* Anyone involved in the project can access the information such as drawings, specification, bills of quantities, letters, etc., with a single source database.
* A project directory stores all the information on parties involved in the project.

- There is a 'template' function for producing all project correspondence including letters, reports, fax's, etc. This ensures the quality of outgoing documents for all projects are maintained.
- Key events for every project such as the start and completion date can be created for monitoring the progress at different milestones.
- Due to the increase in information intensity of the construction process created and used by project members, the company began to realize they could create an environment in which various forms and formats of information could be linked together for easy access and distribution.

Capability Building

The training was conducted throughout the organization including branches, and open to all staff, this consisted of half-day training for all staff to use the basic functions of System Y. Each level of a module is a separate training exercise. When staff successfully complete the basic level of training, they move to the next, more advanced level. Their access to System Y also depends on their training achievement.

The training also conducted for System Y includes

- Series of 18 emails distributed to all staff
- One to one training
  - This is more on an ad-hoc basis and upon request.
  - This is personal one to one training provided to the staff who need more attention about the uses of System Y.
  - The Assistant Training Manager handles this type of session.
- Workshops
  - This is a hands-on workshop about the uses of System Y.
  - The Training/Assistant Manager and Vendor B's trainer conduct this workshop.
  - This workshop is divided into three stages: beginners, intermediate, and advanced.
  - Half day training of all staff to use the basic functions of System Y. Each level of a different module has separate training exercises when the staff successfully complete the basic level of training.
  - All staff were required to pass a lower level training module in order to get to the advanced level. Their access to System Y also depended on their training achievement.
  - Upon completion, the certificates were awarded to the successful trainee.

Training managers give induction courses to new staff, according to the multi-level needs.

The Criteria Relating to the Readiness Model in Level 3 / 4

- Type of skills; considerable technical and project management
- Capability building; centrally integrated training within organization.

Target Status (the Readiness Level that Organizations Desire to be at in the Future)

Type of Skills

Everyone, including the surveyors and administrative staff, expect work under general direction and receive work tasks in the form of project objectives, and are able to plan their own work to meet target costs, time, and quality. Everyone has a good understanding of the relationship between their own area of specialization to other areas. The accounts department is able to understand and appreciate the valuation process.

Capability Building

The existing training methods are expected to be applied for the future and to establish collaboration and knowledge sharing within the company and design team.

The criteria relating to the readiness model in Level 5

- Type of skills; cross disciplinary experience
- Capability building; based on knowledge sharing.

The results of skills analysis are shown in Table 4.

Table 4. People: Skills Analysis

Status	Prior to	At termination/Current	Target	Gap
Maturity Level	1	¾	5	2,3,4,5

Roles and Responsibility of IT Staff

There are misconceptions about who is really responsible for managing IT within Organization BBBBB between the Business System Manager and the IT Manager. The newly appointed IT Manager was appointed at senior management level in the organization structure. However, the size of the IT Department is small and too inexperienced to cope with a high level IT system.

Status Prior to Project

Position of IT/IS Head

The company had an IT Manager with senior management status.

Roles

IT/IS strategy

The Criteria Relating to the Readiness Model in Level 5

- Position of IT/IS Head; IT Manager with senior management status
- Roles; IT strategy.

Current Status of the Project

Position of IT/IS Head

The company has a dedicated IT Manager, but who is still under the Managing Partner. In addition, the Business System Manager is also in a senior management position.

Roles

The role of the IT Manager is to deal with all technical matters in IT/IS including the development of the IT/IS infrastructure (hardware, networking and security) and application.

The IT Manager reports directly to the Managing Partner and the staff of the IT Department consists of 7 employees:

- The IT Manager
- 3 technicians
- 1 programmer/analyst
- 2 trainers

The Criteria Relating to the Readiness Model in Level 5

- Position of IT/IS Head; IT Manager at senior management status
- Roles; IT strategy.

Target Status (the Readiness Level that Organizations Desire to be at in the Future)

Position of IT/IS Head

The position of IT Manager is expected to remain the same in the organizational structure.

Roles

The role of the IT Manager is expected to replace the position of Business System Manager in parallel with the expansion of the IT Department and is to help the senior management align the IT/IS strategy with the business strategy.

The Criteria Relating to the Readiness Model in Level 5/6

- Position of IT/IS Head; IT Manager at senior management status
- Roles; IT/IS strategy and business strategy.

The results of roles and responsibility analysis are shown in Table 5.

User Involvement

The awareness of the importance of user involvement is increasing among the senior management. The users did not participate in the decision making to purchase System Y and only participated in System Y's development.

Table 5. People: Roles and Responsibility of IT Staff Analysis

Status	Prior to	At termination/Current	Target	Gap
Maturity Level	5	5	5,6	6

Status Prior to Project

User Involvement

The role of the users is to define the needs and requirements of the proposed system, and only selected users were consulted to give input. No user group was formed to perform tests and the 'champion' of the proposed system normally made all the major decisions.

The Criteria Relating to the Readiness Model in Level 2

*   User Involvement; Individual consultation.

Current Status of the Project

User Involvement

A group of users have been selected to be a member in the Steering Group, led by the Business System Manager. Members of the steering committee consisted of over twenty staff members from different levels, with a spread of skills and aptitudes, and from all offices (including the partners to administrative staff) of the practice. The role of the users in the Steering Group is the identification of user needs, including functional and non-functional requirements. The Steering Group had one central meeting then were consulted and involved in specific skills areas up to installation, and the committee exists only for the large IT/IS projects and will be dispersed after the completion of the projects.

The Criteria Relating to the Readiness Model in Level 5

*   User Involvement; ad-hoc user group (steering committee).

Target Status (the Readiness Level that Organizations Desire to be at in the Future)

User Involvement

The users are expected to be involved in the systems development by contributing to most of the major decisions, identification of users' needs and requirements, testing the prototype, and approval of the system. However, there is no intention to set up a permanent user group as the practice plan is to go for vendors for full implementation of the future system.

Table 6. People: User Involvement Analysis

Status	Prior to	At termination/Current	Target	Gap
Maturity Level	2	5	5	3,4,5

The Criteria that Relates to the Readiness Model in Level 5

• User Involvement; ad-hoc user group (steering committee).

The results of user involvement analysis are shown in Table 6.

WORK ENVIRONMENT

Organizational Behaviour

Prior to the project, the practice saw IT/IS as a tool for performing their work tasks. The IT system and applications purchased developed more focus on specific groups and individuals, and not on the organization as a whole. IT skills and knowledge were measured through the user's competency at using the system, and the relationship between groups was improved in terms of solving the problem. However, there were gaps between the two main groups of users existing in the practice; the 'younger' group who continuously demand more advanced IT systems, compared to the more 'old fashioned' group which preferred the previous practices and systems and were not welcoming towards the new system.

Status Prior to Project

Characteristics

Everyone had their own way of managing information and used a different approach in performing common work tasks such as storing project information in different files and directories. The dissemination of information within the practice took a long time and was mostly done manually which lead to a non-data sharing culture within the practice. The different work processes used in many workgroups within the practice made it even harder. The process also depended solely on individual efforts, such as how long they kept the document until it was filed away, and also how they managed the filing system. Consequently, the company fell into a crisis when a member of staff left the company.

The possibility of document loss was also high. Members of staff without authority were able to take confidential documents without prior consent. The management saw IT as helping to perform business operations, and not as a strategic use; and saw the performance of IT systems in terms of technology delivery (speed, accuracy and quality).

The Criteria Relating to the Readiness Model in Level 2

- Characteristics;
  - Technology led,
  - IT success measured in IT terms rather than impact made on the business.

Current Status of the Project

Characteristics

The practice started to develop a data sharing culture, and there is an improvement of work co-operation within and between workgroups. Everyone has knowledge of what the others are doing. The practice share processes for common tasks such as creating, storing, retrieving and disseminating information inside and outside the organization – such practices have become more transparent. System Y does not change the way people perform their job but just makes the processes easier. IT success is mainly measured in terms of its efficiency, such as how it can make processes run more smoothly. For example, they measured the benefits of System Y by the number of complaints, less time searching for documents, and fewer filing cabinets (operation efficiency).

The Criteria Relating to the Readiness Model in Level 4

- Characteristics;
  - Sharing processes
  - IT is vital for streamlining business processes.

Target Status (the Readiness Level that Organizations Desire to be at in the Future)

Characteristics

The senior management plan to create an IT system environment towards a customer focus and centralized knowledge. The practice sees IT as one of the vital elements that lead to competitive advantage by producing better products and services for

Table 7. Work Environment: Organization Behaviour Analysis

Status	Prior to	At termination/Current	Target	Gap
Maturity Level	2	4	5	3,4,5

the customer. The practice also encourages more user participation in IT system development in the future and users are expected to become a main reference during IT system development projects. The interaction and relationships between the users and IT staff is improving in parallel with the plan for IT department expansion in the future.

The Criteria Relating to the Readiness Model in Level 5

- Characteristics;
  - IT/IS used to provide better products and services to customers
  - Staff encouraged to input ideas.

The results of organization behaviour analysis are shown in Table 7.

IT Policy

Organization BBBBB centralized all of their IT/IS activities including purchasing and managing the IT infrastructure and applications. Previously, prior to the establishment of the IT department, the quantity surveyor (currently appointed as Business System Manager) controlled the management of IT across the organization. Currently, the IT department controls the IT/IS activities across the organization. However, the IT Manager has to report to the Business System Manager on every decision they make. The Business System Manager is the 'champion' of the IT/IS management across the organization and the management is expected to practice the current policy on managing their IT/IS activities in the future.

Status Prior to Project

IT/IS Activities Control

Mostly controlled by Business System Manager.

The Criteria Relates to the Readiness Model in Level 3

• Centralized IT infrastructure and application policy.

Current Status of the Project

IT/IS Activities Control

The IT department controls all IT/IS activities (centralized IT infrastructure and applications). Every time the business units want particular applications they have to submit applications to the IT Manager, making a business case. If it is not considered expensive, then the IT Manager will discuss with the group leader and then get approval and install. Otherwise, the IT Manager will require approval from the Managing Partner.

The Criteria Relating to the Readiness Model in Level 3

• Centralized IT infrastructure and application policy.

Target Status (the Readiness Level that Organizations Desire to be at in the Future)

IT/IS Activities Control

The IT department controls all IT/IS activities (centralized IT infrastructure and applications).

The Criteria Relating to the Readiness Model in Level 3

• Centralized IT infrastructure and application policy

The results of IT policy analysis are shown in Table 8.

Table 8. Work Environment: IT Policy Analysis

Status	Prior to	At termination/Current	Target	Gap
Maturity Level	3	3	3	Improvement

IT Policy

There was a lack of senior management awareness towards the potential of IT for their organization. Previously, the IT development was organized on an ad-hoc basis. System Y implementation was the first time the IT development in the practice had gone through the proper planning process. The senior management sees IT/IS as tools for smoothing the business processes, and the Business System Manager plays a vital role in representing senior management's views.

Status Prior to Project

Communication

The communication regarding the IT/IS activities was mostly done through manual methods such as through memos, letters and meetings. Most communication between senior management and staff was through the Business System Manager, and the communication between the vendor and the users was also through the Business System Manager. No communication planning existed and everything was done on an ad-hoc basis.

Participation

There was little participation of senior management in IT/IS activities and the senior management were mainly concerned with the IT/IS expenditure.

The Criteria Relating to the Readiness Model in Level 2

- Communication; no established standard for communication (ad-hoc)
- Participation; ad-hoc participation.

Current Status of the Project

Communication

The communication regarding the IT/IS activities including IT/S project is mainly done through emails and during training. Apart from that, communications through memos, letters and meetings is still popular. The senior management began to communicate to the users directly through emails and meetings. The senior management comment on, and answer, any question from the users through email. E-mail played a major role as a communication channel during System Y's development. For example, the introduction and guidelines on using System Y were done through a series of 18 emails. The vendor also conducted an interview with the users to seek out their needs and expectations towards System Y. However, only selected users were chosen to participate in the interview. The practice started to have a communication plan for IT/IS development.

Participation

In addition to the expenditure on IT/IS, the senior management were concerned with the disruption caused during System Y implementation to the business process. The senior management also started to express their concern regarding the improvement/ enhancement of a future System Y to improve client satisfaction, while the Business System Manager participates actively in System Y development.

The Criteria Relating to the Readiness Model in Level 3

- Communication; organization-wide policy and standards for communication
- Participation; Participate only on large scale and high cost IT projects

Target Status (the Readiness Level that Organizations Desire to be at in the Future)

Communication

The practice expected to develop communication planning across the organization for their IT/IS activities, from the experience gained in System Y's development.

Participation

For System Y to successfully satisfy the clients, the senior management have to play a vital role to participate in measuring the performance of the IT/IS project from the strategic perspective.

The Criteria Relating to the Readiness Model in Level 4/5

- Communication; a communication plan is expected for all activities
- Participation; Participation in performance measurement of IT/IS project.

The results of Organization BBBBB behaviour analysis are shown in Table 9.

Table 9. Work Environment: Leadership

Status	Prior to	At termination/Current	Target	Gap
Maturity Level	2	3	4/5	3,4,5

SUMMARY AND FINDINGS

Overall processes for managing information are improving (level 2 to level 5 - process) since the introduction of System Y where;

- Overall processes were not defined and documented prior to the project.
- Also, prior to the project, everyone followed the same process (scope identified) in managing information; however the processes were unable to ensure quality and standards.
- At the current status, the information management began to be standardized and everyone followed the same process which established quality and consistency organization-wide.

Prior to the project, the development of IT/IS was driven to perform daily work tasks that required the company to run a business. Examples of those work tasks are designing, costing, procurement, programming and reporting (level 1 – top management perception). Organization BBBBB then developed System Y to streamline communication organization-wide, which the previous system did not have the capability to do, and to reduce cost for document reproduction (level 2, 3 – top management perception). System Y development is currently continuing to achieve its target for managing the information supply chain within industries (level 5 – top management perception).

The level of IT skills prior to the project was relatively low (level 1 - skills). This was due to the fact that most of the application systems used basic IT functions such as switching the computer on/off, and creating, saving and printing documents. The application use was simple and basic and did not require much IT skill. The daily operations were of applications such as word perfect and spreadsheets for creating letters, documents, correspondence, specifications, project planning, estimating, bills of quantities, etc. The introduction of System Y has helped the company to increase their staff's IT skills. For example, the customization tasks of System Y such as defining functional and non-functional requirements are mostly done by in-house expertise, along with Vendor B's expertise. The in-house IT/IS experts also gained system development experience during System Y's development. The staff's IT skills are expected to improve in the area of using IT for decision making, and across disciplinary skills, parallel with the further enhancement of System Y in the future (level 2,3,4,5 -skills).

Prior to the project, the position of the IT Manager was at senior management level to set up IT/IS strategy. The Business System Manager was also at the senior management position (level 5 – roles and responsibility of IT staff). The position of the IT Manager remained the same during System Y's implementation and then remained so, but the role of the IT department is now to set up IT/IS and business strategy (level 5, 6 - roles and responsibility of IT staff).

There was no improvement in IT/IS activities control since it remained centralized, for all of their IT infrastructure and applications(level 3 – IT policy). In the improved system, the users get involved in the IT/IS development, which is an improvement from the ad-hoc individual participation of the previous system (level 2 – user involvement). They get participation in the steering group led by the Business System Manager in System Y's project (level 5 – user involvement). However, the Steering Groups only exist for the large IT/IS projects, and will be dispersed after the completion of the projects.

Two groups exist within the company. There is a gap between the two main groups of users existing in the practice. They are the 'younger' group which continuously demand a more advanced IT system, compared to the more 'old fashioned' group which prefer the previous or current practice. Prior to System Y's implementation,

Figure 4. Summary of the assessment of System Y's implementation

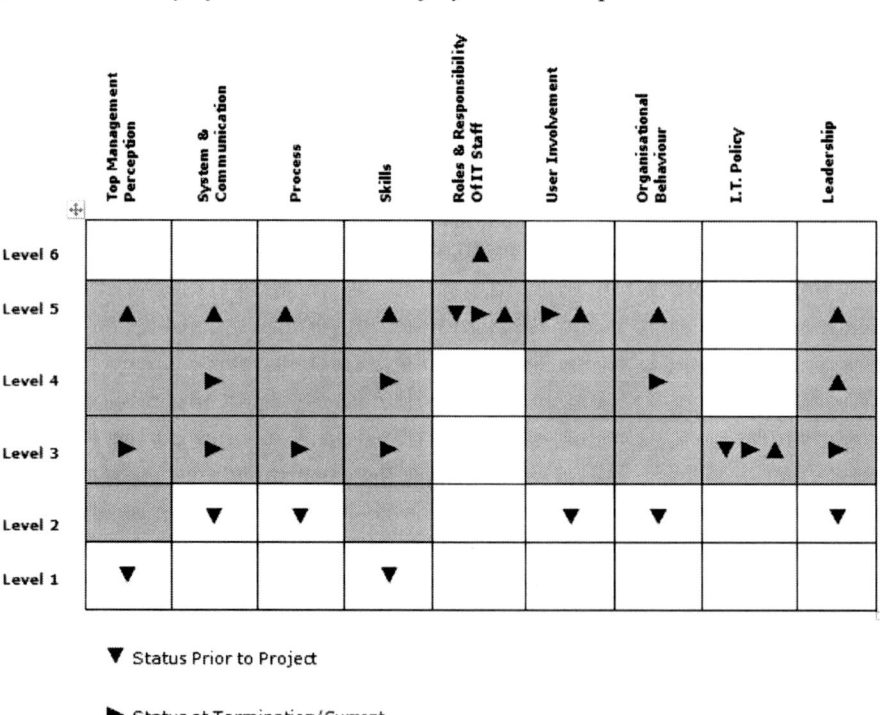

▼ Status Prior to Project

► Status at Termination/Current

▲ Target status (readiness level that organisations desire to be in the future)

[ ] Gap

the management saw the performance of IT systems in terms of technology delivery, i.e. speed, accuracy and quality (level 2 – Organization BBBBB behaviour). However, this perception changed after System Y became 'live'. The management sees IT/IS as vital for smoothing the business process and as a tool for achieving competitive advantage (level 4,5 – Organization BBBBB behaviour).

The summary of the assessment of System Y's implementation according to the readiness model can be found in Figure 4.

CASE STUDY ADOPTED FROM

Salleh, H. (2007), "Measuring Organisational Readiness Prior to IT/IS investments", PhD Thesis, School of the Built Environment, The University of Salford, UK

Chapter 5
Banking for the Future:
Starting Anew

Yasser Al Saleh
University of Salford, UK

Eric Lou
University of Salford, UK

EXECUTIVE SUMMARY

This chapter presents the case of a bank that was established in late 1973 by an initiative from the Government, as a joint venture between the Ministry of Finance, the Central Bank, all commercial banks registered in the country, insurance companies, and some large industrial investment firms. The chapter presents the before and after analysis of IT readiness in the bank. Through the analysis it was realised that the use of IS/IT in the Bank is well behind that normally expected from a bank, which is set up to promote the industrial sector of a major finance centre. The Bank management needs to recognise that technology is changing too fast for the non-specialist to keep up. It is therefore an essential part of a truly professional relationship between a business and its IS/IT staff that they help management to understand that business opportunities which would arise from IS/IT utilization in new areas such as electronic commerce. This vision does not exist in the Bank, at least not in those professionals whose position would give them the necessary influence. This means that the Bank's top management need is to consider IS/IT as a strategic tool to achieve a competitive edge in pursuing the Bank's goals.

DOI: 10.4018/978-1-61350-311-9.ch005

BACKGROUND AND HISTORY

The Bank was established in late 1973 by an initiative from the Government, as a joint venture between the Ministry of Finance, the Central Bank, all commercial banks registered in the country, insurance companies, and some large industrial investment firms. The Bank was established with a share capital of US$35 million, which was subsequently increased to US$70 million, with the Government holding a 53% share. The Government also provided the Bank with a long-term loan of US$700 million, with a grace period of five years and a 2.7% annual rate of interest. In December 2000, Parliament passed a new law allowing the Government to provide the Bank with a loan facility up to US$700 million in the form of a revolving credit. The primary goal of the Bank is to promote industrial development in the country by pursuing the following objectives:

- To participate in developing a long-term strategy for industrial growth and identifying those sectors and activities which would best suit local conditions and constraints.
- To initiate industrial projects and investments in promising sectors.
- To provide financing, whether in the form of equity, or medium or long-term credits for new projects, as well as the expansion of existing ones.
- To finance projects outside the country, with an emphasis on the nearby region, especially where country interests are involved.
- To bring new technologies to the country and identify foreign partners with the necessary expertise.
- To support the development of domestic money and capital markets in cooperation with other major financial institutions.

The penetration of IS/IT into the Bank is low. Out of 250 staff, only 70 are registered users of the mainframe core system. The concentration has been on processing the basic business transactions through a system which was inaugurated in 1981. This resulted in poor design and an ineffective user interface in delivering high quality services to the users.

Sequence of Events

In the second half of 1999, the Bank's top management (Bank's chairman) pushed for "an idea to provide advanced services" to the Bank's customers, because of passing comments made by some members of the Bank's Board of Directors, many of whom are representatives of IT-intensive private sector organizations. Those 'advanced services' refer to Internet banking, telephone banking and branching (there is only the main headquarters currently). When this idea was presented, top management established a committee of IS/IT staff to do a study to determine whether it was possible to improve the current IS/IT to meet those new needs or whether a new system should be implemented. The vendor of the current system was approached for possible improvement of the current system to meet the new needs. The vendor provided a hardware solution only.

This study gave both users and IS/IT people the opportunity to voice their dissatisfaction with the 17-year-old system. The Bank's study committee was faced with two options, either to change the hardware only and mainly meet the Chairman's ideas, or change the whole system and meet everybody's needs i.e. Chairman, IS/IT unit, and users. The study recommended the latter option. "To have an objective outside opinion, the Bank hired a consultant firm to conduct a study", which gave recommendations regarding "the best way to tackle the solution." The consultant's study suggested replacing the current system with a new one. It suggested doing that in small steps until the Bank reached the needed state in meeting the new requirements. First, to install a bank-wide network and implement some office automation software based on this network. This software should include financial software and internal e-mail. Secondly, to train the users to use those tools and to share files and printers, followed by the Bank installing a new and total core banking system. After completing the previous steps, the Bank can install on top of the new infrastructure whatever tools the Bank requires in order to support the Bank's needs i.e. Internet banking, telephone banking, etc. The consultant has also recommended that the Bank should hire a consultant company to study the requirements and produce a Request For Proposal (RFP) for the core system. This consultant should help in the selection process of vendors and systems. The Bank started to implement the consultant's recommendations by performing steps described in the following paragraphs.

In December 1999 the Bank signed a contract installing a new network. In April 2000 it was ready to be used with office automation and e-mail software. Even though the new core system(s) had not yet been decided upon, the choice of the network configuration was made. The project leader explained that he implemented a flexible and almost standard configuration that would support whatever candidate core system the Bank might implement later. This would include Client/server, TCP/IP and Ethernet, Relational DBMS etc. Users were trained in using the network features of file and printer sharing, office automation software, e-mail, and Internet browsers. They started to use the new tools and centralized backups were periodically taken by IS/IT unit. Efforts to discourage users from using the old unconnected small PC-based systems that they built and kept over the years were not successful in most cases. The bank's IS/IT personnel were also trained in using the functions needed for operating the tools that they recently acquired, such as security and software monitoring. The old system was still running separately from the new network and tools, where some users had two terminals on their desks.

In May 2000, the Bank chose an international consulting firm to be the consultant to implement step 3 of the recommendations made by the first consultant's study. This new consultant firm would also help in the selection process of vendors and systems. Besides choosing this consulting firm, five other well known international consulting firms were proposed for this contract. Among the criteria that the Bank had for choosing the consultant was the degree of experience and qualifications of the people the firm was going to assign to the project. Résumés were supplied by two of those firms, while others refused. It was specified in the contract that it was forbidden for the consultant firm or any of its subsidiaries to enter in any bid for the system(s) under study.

The contract was signed with the consultant firm in May 2000 to start the consultancy services. Requirement gathering and assessments from the Bank's departments then produced an RFP that specified all the department requirements for a new system, which started a month later. Each department reviewed and signed the document produced by the consultant. This activity took about two months. The IT steering committee meeting approved the proposed RFP. This committee consisted of the deputy bank chairman, head of the legal affairs unit, and the project leader from the Bank. In late August 2000, five international vendors were invited to tender. The consultant then evaluated vendors' responses to the RFP by a scoring system. One of the vendors did not bid because it did not think its system could meet the Bank's requirements. Another vendor scored very low. The summary of the overall scoring is shown in Table 1.

After a long and tedious selection process, all studies and evaluations favored either V-B's proposal or the combination of V-C with V-F proposals, but the steering committee selected V-A (V-F did not enter the bid, since it was for a total solution banking system and V-F had a financial system only). After it was selected, V-A conducted "Gap Analysis" which took approximately 4 weeks. In the Gap Analysis, both the Bank and V-A identified all the gaps between how the Bank did business and how the software did business, and if there were differences, how they could be resolved. After the Gap Analysis, V-A produced the final cost for all the customization agreed upon. Furthermore, the final solution cost was reached, and changes were clearly labeled to be: implementation tasks, system customization,

Table 1. Results of vendor responses to the proposed RFP

Vendor A (V-A)	Vendor B (V-B)	Vendor C (V-C)	Vendor E (V-C)
84.13%	94.48%	84.02%	69.89%

Bank policy and procedure change, or any work-around changes to be done to meet the Bank's requirements. V-A started the customization of its proposed Core Banking System according to the agreement with the Bank. During this process, there were some amendments and changes, which arose for different reasons. New government regulations needed to be implemented in the system; also, as the users started to be subjected to the environment of development, they started to have new ideas and requirements. Most of them were minor, but some had a major impact on the customization. This was a cause of some friction both internally between the IS/IT unit and users, and externally between the Bank and V-A. As the customization process was going on, the Bank implemented a training programme for both users and IS/IT staff. Up until the time this research ended, the implementation of the core banking system had not yet been completed.

ANALYSIS AND DISCUSSION

The following sections address each attribute of the four domains of the readiness/ GP model within the context of the Bank.

IT INFRASTRUCTURE

Systems and Communication

Investment in IS/IT has been static over a number of years, contributing to the system portfolio being severely out of date. In a report published in 1999, it was documented that almost all system changes were reports modifications. This has followed a year in which no application development was done "due to the Y2K". The limitations of the mainframe system have led to several departments re-entering information from printouts and other sources into (stand-alone) PC-based worksheets – to enable information analysis. Where individual PCs are in use, there is no means to ensure that compatibility is maintained by using the latest version of the software across all users. There were also a high number of special report requests and report changes by the users which contributed to the IS/IT unit performance problem. The Bank's operational processes were riddled with apparent reconciliation errors that arose because the data flows between systems was poorly designed as in loans, LC (Letter of Credit), LG (Letter of Guarantee), and closure-of-facility. There was no clear distinction between operational systems and informational systems. This lead to performance problems as updated transactions had to fight for database resources with complex queries. Given the number of reports requested by the users this issue could have contributed to the performance problems. The Bank installed a network to be used with office automation and e-mail software. The network configuration was flexible and almost had a standard configuration that could support the core system the Bank is implementing. This would include Client/server, TCP/IP and Ethernet, Relational DBMS etc. The network enabled users to share files and printers. It also provided them with office automation software, e-mail, and Internet browsers. Users started to use the new tools and the central IS/IT unit periodically took centralized backups. There were serious efforts towards convincing users to discard their old, small, PC-based, unconnected, almost individually-based systems. These efforts have not proven fruitful in most cases. On the IS/IT technical side, new features were provided with the network, such as security, automatic periodical backup, and software monitoring. The old system was still running separately from the new application tools that are running on the communication network, where some users have two terminals on their desks.

Status Prior to the Project

Formally, the Bank shows some of the level 3 system characteristics for the situation prior to the project, but because the system became technologically obsolete at the beginning of the project, it actually agrees with the general description of mostly level 1 and partly level 2. Even though the Bank has had a centralized system for many years, management allowed the users to build uncontrolled, unconnected small systems in an ad hoc fashion, where all are in the financial area. Also, because of the many errors in the old systems, the IS/IT unit was overloaded with requests for fixes. The situation proved that the Bank situation went backwards in the maturity progress. The following is the description of level 1 and 2 in the readiness/GP model.

Level 1

- Almost all existing systems are small packages for financial operations
- Ad hoc IS/IT development where each unit invests independently from the rest of the organization and the approval process of IS/IT projects differs between units
- Information systems are independent and unconnected, organization-wide or even within the same group, which makes IS/IT portfolios of each group differ from the rest of the organization
- IS/IT development, maintenance, implementation and training decisions are made at the group level, where groups manage their own IS/IT resources according to each group's needs, in isolation from the rest of the organization.

Level 2

- An increase in the number of IS/IT application systems is being developed or purchased, but concentration is still on operational systems in the financial area, while a small number of other core business-oriented systems are being developed.
- Many of the IS/IT application systems still overlap in purpose, function and data stored in the organization, where only some hardware, system software, and possibly a network are shared between groups
- A large maintenance load is being placed on the IS/IT function because of the ad hoc nature of most systems
- All data are stored in units' systems, except data needed for organizational reporting, which are transferred to central systems.

Current and Target Status

The current situation shows many of the characteristics of the target systems level. The existence of an organization-wide network and shared integrated office auto-mation applications where security, backups, and software monitoring are done centrally, are all characteristics of systems' level 4. The new system, as proposed by the consultant, agrees with the general description of level 4 (from the interview with the project manager). The description of this level is as follows:

- All required operational IS/IT is mostly in place and some DSS might start to appear
- Office automation is integrated and unified/standardized organization-wide
- Existence of an organization-wide network, where all groups are connected and the central IS/IT function provides communication services for all groups in the organization
- Central coordination in the use of IS/IT throughout the organization, where an effort is made by groups' IS/IT functions to follow standards set centrally
- The organization-wide network is starting to be utilized to connect users to whatever shared applications and information systems that are needed
- Extensive use of standard e-mail messages throughout the organization, and there is evidence of dependence on the organization-wide network to conduct formal communication

The results of systems analysis are shown in Table 2.

Table 2. IT Infrastructure: Systems Analysis

Status	Prior to	Current	Target	Gap
Maturity level	1-2	4	4	-

PEOPLE

User Involvement

Many of the staff in the IS/IT unit are not able to maintain skills consistent with today's needs. Many key skills are missing and technical skills, such as programming and systems analysis skills, are stuck in the past, which makes the underlying assumed skills of the job titles these staff hold differ from the skills they actually have. For many years, the IS/IT staff lacked the business skills to assist the users to define their needs and to develop new features in the system. In agreement with the new organizational chart (see Figure 1) a Database Administrator (DBA) was hired. The DBA's task will be the administration of the proposed relational DBMS.

Figure 1. Current IS/IT organizational chart

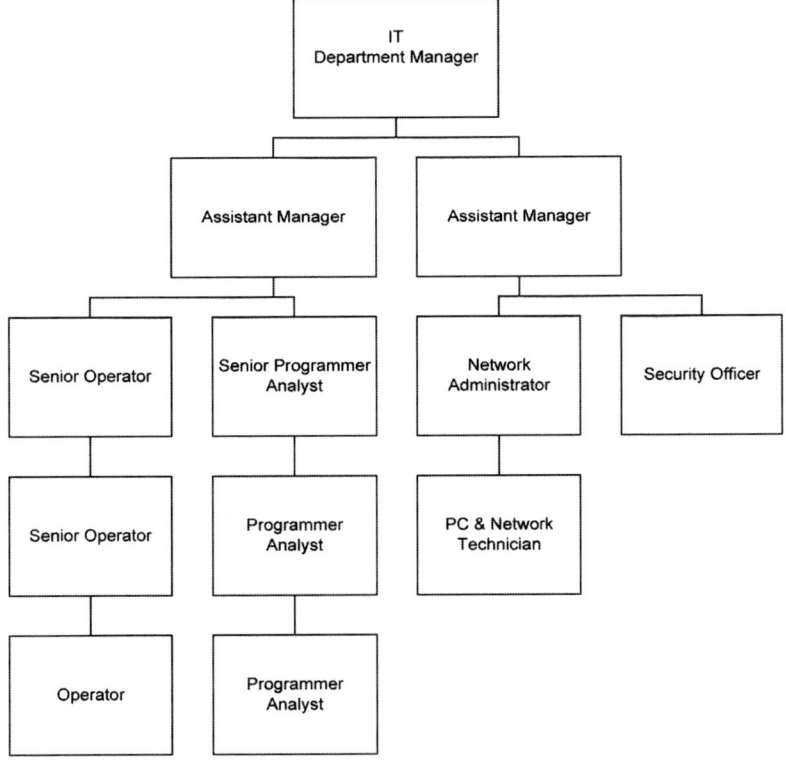

Following the implementation of the new system, the maintenance operations will be outsourced to the vendors of the new hardware, systems' software, and core banking system. This is reflected in the new organizational chart of the IS/IT unit (Compare Figures 1 and 2). The new organization does not contain any programmer, and system or business analyst positions.

There is a need for V-A to appoint a credible business analyst(s) to work as a liaison point between users and the IS/IT unit, for implementing changes and new requirements in the system. This might be a potential problem area if the vendor does not provide adequate expertise and resources, a problem that is common in the country's IS/IT sector. For the Bank to totally restructure the IS/IT unit, relying almost entirely on the vendor's support, is like wandering into a minefield, especially in a market place that is known for having weak after-sales support, and even more so with a vendor, such as V-A, which has no prior experience.

Status Prior to the Project

Even though staff show some of the level 2 staff characteristics in the readiness model, in reality, the underlying skill and experience assumptions of the job titles the staff hold do not reflect the actual staff situation. This is because of the obsolete

Figure 2. Proposed IS/IT organizational chart

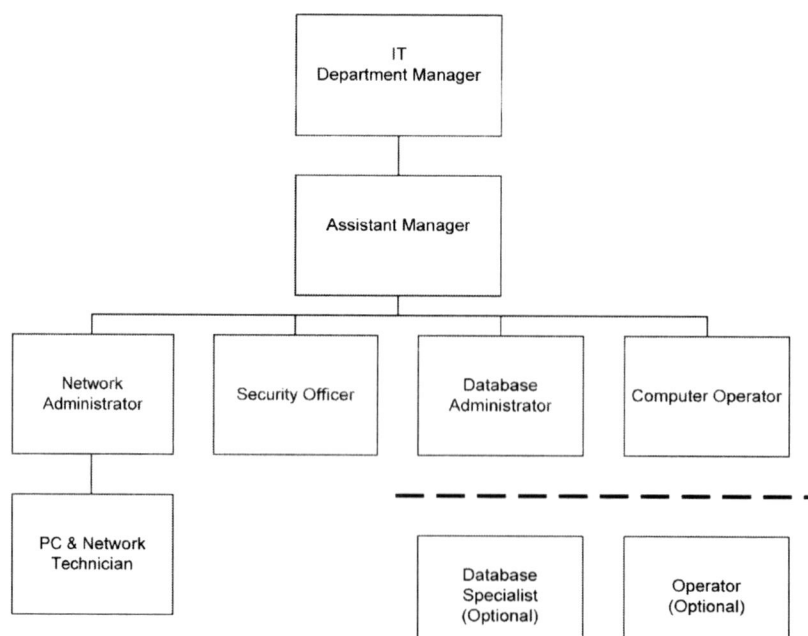

Table 3. Old/current numbers and description of IS/IT staff

IT Management	3
Business Analysts/Developers	3
Network Support and Security	3
Computer Operations	3
Secretarial Work	1
	13

skills of those staff. In addition, only one staff member (IS/IT head) has the technical know-how. Instead of assuming the managerial role, he was working as programmer, technician, and analyst, and so the IS/IT department and the system was left in a poor condition. Although the situation could have been interpreted at level 2, in reality level 2 staff criteria did not apply to the Bank situation at the beginning of the project. The staff situation is more in line with the general description of the staff characteristics of level 1 in the readiness model. The description of this level in the readiness model is as follows:

- No dedicated IS/IT staff or small number of low-level technicians and programmers
- No manager allocated responsibility for IS/IT
- External contractors may be used to develop/install systems as required
- Users' new recruits are not expected to have IS/IT-related skill

Current Status

As the new organizational structure of the IS/IT unit has been implemented and many of the required staff for the new system have been recruited, the current/proposed staff situation is in line with the general description of the staff characteristics of level 3 in the readiness model, providing the vendor will supply the technical staff and expertise necessary for the smooth operation of the system. This level's description is as follows:

- Added to the programmers and analysts, dedicated IS/IT planners and database administrators are appointed
- Almost all needed technical specialist staff are in-house
- A technically-oriented IS/IT manager is appointed or DP manager might have a change in title

Table 4. Consultant's proposed IS/IT staff numbers and description

IT Management	2
Database Administration	1
Network Support and Security	3
Computer Operation	1
Secretarial Work	1
	8

- IS/IT workforce are coordinated with current and future IS/IT needs at both the organizational and unit levels
- New user recruits are expected to have specific IS/IT-related skill

Target Status

The new system will span all organizational functions – the Bank needs to have a dedicated liaison body of staff with adequate understanding of both the different business functions and the functional capability of the Bank's IS/IT. Also, because of this wide span, the managerial position of the IS/IT unit heads need to be of senior calibre in order to have the required power and respect within the organization. The target level of staff should agree with the general description of level 4 staff characteristics in the readiness model. The level 4 description is as follows:

- In addition to the programmers, systems analysts, and data base administrators, business analysts now exist to act as a liaison between their units and the IS/IT unit
- High level manager for IS/IT services area is appointed, with middle management status
- Organizational staff (IS/IT and user) performance is quantified and measured against quantitative performance baselines

The results of staff analysis are shown in Table 5.

Table 5. People: Staff Analysis

Status	Prior to	Current/proposed	Target	Gap
Maturity level	1	1	3	3

Skill

The IS/IT unit have out-dated skills and it would require significant training effort to bring them up to date. Many skills have become obsolete, such as programming and analysis. The pace of change in IS/IT has been dramatic over the last 17 years, since the time the old system was inaugurated, and the Bank has not kept up. Users complained of the lack of banking business knowledge among the IS/IT staff; and of the lack of any kind of IS/IT skills that are of value compared with the current technological situation in the outside world. The training programme introduced after the installation of the network and office automation software seems to enhance the IS/IT related skills of users and has enhanced the skills of mainly the technicians among the IS/IT staff. The staff chosen to participate in the training programmes were assigned operational responsibilities in running the new network and software. This was coupled with promises of new organization in the IS/IT department, change of titles, and promotions, once the new system's implementation was completed. The new project did not contribute in any serious way to the skill of the majority of IS/IT unit people because the external consultant staff performed most of the studies and development/customisation. This made many staff members insecure and prepared to leave the Bank.

Status Prior to Project

Even though skills show some of the level 2 characteristics in the readiness/GP model, this does not in reality reflect the underlying assumptions of the job titles staff hold because of the obsolete old skills. Also, because the system's technical know-how is limited to one person in the IS/IT unit, level 2 criteria do not apply to the situation prior to the project. So, at the beginning of the project, the status was most in line with the general description of level 1. The description of this level in the readiness/GP model is as follows:

- Users find it hard to acquire the skills to use the few IS/IT applications that exist and skills are individually based and jealously guarded from others
- The emphasis is on technology rather than organizational or informational issues, where there are very limited technical skills in the organization/unit as a whole IS/IT skills are specific to individual IS/IT applications
- Required skills are of low level technical nature and there may exist very limited advanced (programming or systems analysis) skills in the organization
- There is almost no IS/IT training provided by the organization.

Current Status

The training programmes were introduced to both user and selected IS/IT staff, the skill situation has currently shown some improvement from the beginning of the project. It now is mostly in line with the general description of level 2 characteristics. The level 2 description is as follows:

- Users begin to have the required training and skills to use the new IS/IT that is being developed/purchased
- Still little in-house technical expertise in IS/IT development (methodology, structured techniques) and other important skills
- IS/IT staff acquire the skills needed to develop and maintain complete systems such as programming and analysis, in addition to being able to install off-the-shelf, ready-made packages
- Limited project management skills
- IS/IT individuals have the skills required to perform their assignments and begin to have the relevant training and development opportunities
- Individuals in the organizational workforce have remuneration and benefits based on their contribution and value to the organization.

Target Status

The new system will span all functions and aspects of the Bank's operations, communication skills are needed for all workgroups, including IS/IT, to be able to complement and actually integrate their work, making the best of the integrated work environment that is facilitated by the new system. Also, new technology introduced in the Bank requires a high level of knowledge and competence to operate and maintain it. Even though the Bank has not decided yet if it will maintain the software itself or outsource this side of the IS/IT unit, the operation of the network, hardware and system software will remain an in-house function, mainly for security reasons. These requirements call for the target level of skill to agree with the general description of level 3 characteristics in the readiness model, which are as follows:

- Considerable technical competence in the organization because of the well developed IS/IT related skills (programming, analysis, security, networking etc.)
- The organizational workforce are constantly enhancing their IS/IT capabilities to perform their assigned tasks and responsibilities
- DP/IS/IT manager and staff lack, but work on building, interpersonal skills

Table 6. People: Skills Analysis

Status	Prior to	Current	Target	Gap
Maturity level	1	2	3	3

- Project management is realized to be needed in this stage, which results in well developed project management skills.

The results of skill analysis are shown in Table 6.

Head of IS/IT

Authority and control of the development and daily activities are concentrated in one key individual (Head of IS/IT). This has negative implications where the security of the Bank is largely vested in that one individual. He appears to be the only person who thoroughly understands the old system, and is also responsible for the quality assurance of the unit output. There is no review body to monitor the quality of the deliverables. The IS/IT unit head did not provide advice to help top management to understand what strategic business opportunities would arise in the future from new technologies. Table 7 shows the new specification for IS/IT unit head as stated in the new IS/IT unit structure document.

Status Prior to the Project

The head of IS/IT did not assume the managerial role, he was performing operational tasks. Even though his official situation agrees with level 2 in the readiness/GP model, in practice the head of IS/IT criteria for the Bank situation at the beginning of the project agree mostly with the general description of the characteristics of level 1 in the readiness model. Level 1 description is as follows:

- There is no individual responsible for systems.

Table 7. Proposed responsibilities of IS/IT head

Proposed responsibilities of IS/IT Head
• Report to the General Manager of the Bank. • Manage and lead the IT staff • Attend the different committees to contribute in the decision making process of the Bank and define the role of IT in implementing the strategies and resolutions. • Prepare the Department strategy in tandem with the overall plans and directives of the Bank.

Proposed Status

The situation that is proposed by the consultant in the project plan (Table 7) for the IS/IT head is to agree mostly with the description of level 4 in the readiness model. Level 4 description is as follows:

- There is an IS/IT manager who has middle manager status.

Target Status

As the new project is considered to be a total solution that spans all aspects of the Bank's operations, the head of IS/IT need to have a senior position. At the time of analysis most of the IT/IS staff were in the lower level management and the top management did not have IT/IS aware members. Therefore, there was a need for transition and either training some top management people on IT/IS or get someone from outside who is from IT/IS background to join the board. The required target and the possible attainable level of IS/IT head of the Bank is level 5 of the readiness model. The level 5 description is as follows:

- There is an IS/IT manager who has senior manager status.

    The results of Head of IS/IT analysis are shown in Table 8.

PROCESS

Practices

The Quality Assurance of IS/IT unit's service, as an example of the process situation, is carried out by the head of the function. There is no independent review body to monitor the quality of the deliverables. There is no quality manual that defines not only the process to be followed, but also the measures that can be used to judge

Table 8. People: Head of IS/IT Analysis

Status	Prior to	Proposed	Target	Gap
Maturity level	1	4	5	2-5

how successful and effective the process is. The Bank therefore does not have a true picture of the cost of the IS/IT process versus the benefits that its systems deliver. There is no clear distinction between operational systems and informational systems. This leads to performance problems as update transactions fight for database resources with complex queries. Given the number of reports requested by the users this issue may contribute to the performance problems.

Status Prior to the Project

The status of the processes fits level 3-GG2, but in a weak way which is almost level 2-GG1.

Current/ Proposed Status

No process change has yet been made, but there are suggestions for customizations of process to match the software. The consultant firm's suggestion on what it terms "BPR" is stated in its Consultancy Proposal to the Bank under the heading "Business Process Re-engineering (BPR)":

"Once a vendor has been selected and a contract negotiated and signed, the team comprised of (consultant firm name), the selected vendor and the Bank will conduct a study to examine what are the recommended changes to the way the Bank does business. The reason for this exercise to be conducted at this stage is to ensure that all changes made are in line with the way the selected software supports the business. The aim is to get the Bank to follow the software and not the other way around. This should minimize the customization cost."

Target Status

The new process level, as the project leader thinks, needs to be in level 4-GG3. In order to reach this level there is a need for an expert in banking processes.

The results of Process analysis are shown in Table 9.

Table 9. Process: Practices Analysis

Respondent:	Prior to	Current	Target	Gap
Stage	Weak 3-GG2	Weak 3-GG2	4-GG3	3-GG2, 4-GG3

ENVIRONMENT

Leadership

The Bank Chairman was instrumental in changing the Bank's system, as he recognized the current 'state of play' at other organizations. This came about after the Bank started to have the idea of expanding. Up to this point, top management had no concern regarding the importance of IS/IT, as it is clear from their declining to invest in IS/IT for 17 years. In an interview with a newly appointed senior manager, he described some of the business objectives for the Bank, operating in what he understands as a changing world. He sees that the Bank needs flexibility to move in any direction that it may wish. For example, retail banking, branches, insurance, remote banking or ATMs. Mixed signals come from top management regarding the way they consider IS/IT. On occasions, they spoke of "computers" being the specialty of the "computer department" and if it had any suggestions they would be willing to look into it. Some suggestions went to top management from junior staff over the past few years, but they were not taken up because "it cost too much and we don't need it that much". There was also pressure from some Board Members to modernize the Bank.

Status Prior to the Project

The level of leadership maturity of the top management before the development of the system and at the beginning of the project fluctuates between the general description of level 1 and level 2 characteristics in the readiness model. The situation mostly agreed with the following general description of leadership:

Level 1

- Has little concern for the potential utility of IS/IT

Level 2

- Considers IS/IT to be the concern of technologists not management but supportive of IS/IT, where priority and thrust are to minimize the expense of IS/IT utilization.

Current/ Proposed Status

No evidence yet that the current level of leadership maturity regarding IS/IT has improved. The change in attitude could be attributed to the pressure from the board of directors. The current situation of leadership IS/IT maturity remains at levels 1 and 2 as described in the readiness model.

Target Status

The Bank management needs to recognize that technology is changing too fast for the non-specialist to keep up. It is therefore an essential part of a truly professional relationship between a business and its IS/IT staff that they help management to understand that business opportunities would arise from IS/IT utilization in new areas such as electronic commerce. This vision does not exist in the Bank, at least not in those professionals whose position would give them the necessary influence. This means that the Bank's top management need to consider IS/IT as a strategic tool to achieve a competitive edge in pursuing the Bank's goals. It also needs to view the flexibility of strategic IS as an asset for the Bank. This means that the required target level of maturity of the Bank's leadership regarding IS/IT is to agree with the general description of level 5 in the readiness model. The description of this level is as follows:

- Considers IS/IT as one of the vital parts of the competitive strategy, where a flexible IT infrastructure is perceived as an asset for competitive edge and is brought up in this way during project justification.

The results of leadership analysis are shown in Table 10.

Culture

The users were complaining of many day-to-day problems and frustrations with the systems. In at least one instance, when an outsider brought this to the attention of the IS/IT unit staff, it was immediately identified as a bug and remedial action was set in motion. This points to a lack of communication between IS/IT unit and users.

Table 10. Environment: Leadership Analysis

Status	Prior to	Current	Target	Gap
Maturity level	1-2	1-2	5	3,4,5

There are no facilities for users to report problems and have them formally recorded. Consequently, there is no concerted plan for remedial action and user frustration builds up, and the trust between user and the IS/IT unit declines. There was also no communications infrastructure for the sharing of information. Users did not have the ability to access information and share it with colleagues and customers. It also severely limited the IS/IT unit's options when considering new systems development. The IS/IT unit believes that the Bank is different from any other Bank and has a unique set of processing requirements. The users however thought that the Bank was not unique and followed normal banking practices, albeit without some of the activities that other Banks perform. In addition, power is concentrated in few individuals; knowledge of the old mainframe system and the whole of the control of the development and day-to-day IS/IT activities are held by a small number of individuals (often only one). The security of the Bank is largely vested in the one individual who appears to be the only person who thoroughly understands the systems.

Status Prior to Project

At the beginning of the project, the status was mostly in line with the general description of level 1. The description of this level in the readiness/GP model is as follows:

- The relationship between user and the limited low-level technicians that may exist in the organization or contracted from outside is of support for the existing IS/IT products that are mainly off-the-shelf, ready-made packages.
- There is no recognition in the organization of the importance of working towards building a constructive relationship between the IS/IT function and users.

Current Status

The current situation of the culture within the Bank is still the same; no effort was made to change it. It is still in line with level 1.

Target Status

Because the new system spans all of the Bank's operations and functions, complementary work is expected between different groups, including IS/IT. For integration to happen, people need to communicate and cooperate. The needed target situation should agree with the general description of level 4 characteristics in the readiness model:

- IS/IT function supports the activities of users.
- There exists an emphasis on organizational integration between workgroups among which is the IS/IT function.
- Workgroups (IS/IT and users) have the responsibility and authority for determining how to conduct their business activities most effectively.
- An improvement in the efficiency and quality of interdependent work, resulted from the integration of the capabilities and knowledge of different workgroups on both IS/IT function and user sides, and with each other.

The results of culture analysis are shown in Table 11.

Structure

Power is concentrated in a few individuals within the IS/IT unit, with a centralization approach where even the most trivial change to a user report must go through a lengthy and time-consuming process. Top management expressed the need for the new system to be flexible to support whatever future business direction the Bank chooses. This requires a flexible structure that combines centralization and decentralization in forming coalitions between the IS/IT unit and users.

Status Prior to the Project

The IS/IT unit situation did not change over the 17 years since the introduction of computers into the Bank. Even though the IS/IT unit is not new it still had the same role from the time it was formed. In addition, users had lost trust in the old system, so they relied on their own separate PC-based systems for many of their needs. This makes the level of structure at the beginning of the project in line with the general description of level 2 characteristics in the readiness model. The description of this level in the readiness model is as follows:

- Separate DP/IS/IT function has recently been introduced where groups are encouraged to seek advice from this newly formed central IS/IT function.

Table 11. Environment: Culture Analysis

Status	Prior to	Current	Target	Gap
Maturity level	1	1	4	2,3,4

- There is a decentralized responsibility for IS/IT function, where groups have full freedom in managing their IS/IT with increased self-reliance regarding IS/IT matters which is apparent throughout the organization.

Current/ Proposed Status

The proposed structure after the implementation of the new system will keep the IS/IT unit in its organizational position. If this happened, then at best, the level of structure would be in line with the general description of level 3 in the readiness model. The description of this level in the readiness model is as follows:

- Official power is vested in the IS/IT function, where a new technical IS/IT manager is appointed or the DP manager might have a change in title to IS/IT manager which goes with a similar change in department name.
- There is an organization-wide IS/IT architecture policy, and standards for telecommunications, preferred suppliers, e-mail, etc.
- Management of the IS/IT function is centralized.
- IS/IT staff seek control of IS/IT matters.

Target Status

Because the new system spans all the Bank's functions and departments, the new system target needs the structure to generally agree with level 4 description in the readiness model. The description of this level in the readiness model is as follows:

- IS/IT function is well established and its mission is to exploit the IS/IT for business purpose and provide competitive IS/IT in a partnership environment with users.
- Decentralized responsibility of IS/IT services with central standards and policy for coordination, implementation and utility.
- Units' IS/IT function reports to units' business manager.
- Significant degree of involvement of users in IS/IT related decisions, where IS/IT investments are derived from users' stated needs.

The results of structure analysis are shown in Table 12.

Table 12. Environment: Structure Analysis

Status	Prior to	Proposed	Target	Gap
Maturity level	2	3	4	4

SUMMARY AND FINDINGS

The use of IS/IT in the Bank is well behind that normally expected from a bank, which is set up to promote the industrial sector of a major finance centre.

The Bank management needs to recognise that technology is changing too fast for the non-specialist to keep up. It is therefore an essential part of a truly professional relationship between a business and its IS/IT staff that they help management to understand that business opportunities which would arise from IS/IT utilization in new areas such as electronic commerce. This vision does not exist in the Bank, at least not in those professionals whose position would give them the necessary influence. This means that the Bank's top management need to consider IS/IT as a strategic tool to achieve a competitive edge in pursuing the Bank's goals. It also needs to view the flexibility of strategic IS as an asset for the Bank. There is no independent review body to monitor the quality of the deliverables. There is no quality manual that defines not only the process to be followed but also the measures that can be used to judge how successful and effective the process is. The Bank therefore does not have a true picture of the cost of the IS/IT process versus the benefits that its systems deliver.

The capabilities of the old/current core system applications do limit the operational capability of the Bank. However, there is a much more serious problem to be addressed. The current IS/IT very poorly serves the Bank's decision makers, at all levels. The existence of large amounts of data on hardcopy listings is a major inefficiency. Users cannot directly extract the information they need from the system; rather, they have used different versions from different PC-based tools to develop many spreadsheets etc. This is an error-prone and time-consuming task that could be eliminated. They should be concentrating on their core competence as a Lending Officer or Credit Assessor, etc., not on developing complex computer literacy skills. If two users take the same data and process them to draw out comparable information, they will often disagree on the results, leading to time-wasting discussion and investigations. There are two reasons for this: first, transcription errors, and second, differences in the analysis they are applying. There is no clear distinction between operational systems and information systems. This leads to performance problems as update transactions fight for database resources with complex queries. Given the number of reports requested by the users, this issue may contribute to the performance problems. However, there is a more serious problem. The Bank's operational processes are riddled with apparent reconciliation errors that arise because the data flows between systems are poorly designed. These scenarios apply equally to loans, Letters of Credit (LC) and Letters of Guaranttee (LG). The closure of a facility has similar problems.

There is no communications infrastructure for the sharing of information. This is probably the most serious consequence of the Bank's failure to invest in modern technology. Without the ability to access information and share it with colleagues and others,within and outside the Bank (such as customers), the users cannot do their job effectively. It also severely limits the IS/IT unit's options when considering new systems development options.

Power is concentrated in a few individuals. Knowledge of the old mainframe system and the whole of the control of the development and day-to-day IS/IT activities is held by a small number of individuals (often only one). This has many implications, one of which is that the security of the Bank is largely vested in that one individual and he appears to be the only person who thoroughly understands the systems.

There is a danger that the vendor will not fulfil its responsibility of providing the needed expertise and other resources for the operation. If this happens, the Bank's staff will be in level 1 instead of being in level 3, as planned by the Bank.

Even if the Bank succeeds in achieving level 3, it is not enough. There needs to be a dedicated and competent liaison body between users and the IS/IT unit.

Since the new project is considered to be a total solution which spans all aspects of the Bank's operations, the Head of IS/IT should have a senior managerial positionAlso, no person in the Bank exists that has the skill and knowledge to take on such a senior position and responsibility. The current project leader was a young junior staff member in the IS/IT department who had no prior experience until he was given the role of leadership of the project on the Bank side. If he is to be the IS/IT Head after the implementation of the project, this attribute (Head of IS/IT) might become a potential problem area. It is essential that the IS/IT unit help top management to understand what business opportunities for strategic advantage will arise in the future from new technologies. This does not currently exist in the Bank.

In general, the Bank shows the criteria of a certain level, but because of the bad management it also shows criteria of a lower level, i.e., point 3 in Stage 3. Even though the Bank had a centralised system from 17 years ago, the old management made the user build uncontrolled unconnected, small systems. It was a step backwards in the maturity progress. It could be that the easiest progress could be made on the systems attribute. To achieve such progress, the organization needs only to buy the IS/IT and networks.

The Bank seems to have taken some correct pre-steps, especially in terms of the IT Infrastructure factor. Currently, at the start of the implementation of the new systems, the Bank already shows that the IT infrastructure attributes have already reached the 'needed' level. The people attributes, and especially culture, could present trouble areas. Culture has not been addressed at all in any pre-steps action. Skills, and especially user skill, could be a reason for delay in the success of the project, since the gap between the 'before' and the 'needed' represents two levels which require lengthy and concentrated training. Also, time is needed for users to gain experience with the new system, which training does not provide. Some level 3 criteria are linked to culture issues especially communication skills, and since culture has not been addressed then this might remain a deficiency.

The process remains an unclear area for prediction. Even though the Bank addressed BPR, it seems that the process changes are connected more to the software package capabilities, features and requirements, than to business needs. Also, it still remains unclear how the user will react to those process changes, especially since they were mainly suggested by the technical people on the Bank, the consultant, and V-A sides. The user was only consulted and was not a participant in, or an originator of, the changes in processes. Also, those changes suggested by the user concentrated on changes in the software, not the process, which could indicate that the user would like to keep processes the way they are. In addition, process changes came late in the project, after the system had already been decided upon, and as part of customisation. This indicates that those changes are more linked to the software capabilities than to business needs.

Project Manager comments regarding the readiness/GP model:

"Many of the aspects in the model are obvious, others are not, but all need to be brought to attention. The model does this in a nice way."

CASE STUDY ADOPTED FROM

Al Saleh, Y. (2002), " IS/IT Success and Evaluation: A General Practitioner Model", PhD Thesis, School of the Built Environment, The University of Salford.

Chapter 6
E-Readiness in Governmental Public Service Institution:
Lessons Learnt

Yasser Al Saleh
University of Salford, UK

Mohammed Arif
University of Salford, UK

EXECUTIVE SUMMARY

This case study revolves around a governmental public service institution, which receives public and government money that it invests. There were several challenges associated with the implementation of the IT system to improve public service. It was found that the organizations need, in the contract, to have the qualifications of the vendor's staff, and agree that prior approval for any change of staff or new recruitment would be agreed beforehand. This is because the vendor's staff had a high turnover. Experienced staff, which were agreed upon by the organization, were assigned to the project for a short time, only at the beginning of the project. The lack of positive relationships between different groups in the organization caused resistance to the required changes in structure and processes. Because key staff considered keeping knowledge and experience to themselves as a job security tool, they were not forthcoming in cooperating with the project team. This was complicated by the almost complete absence of systems' documentation, and the little documentation that did exist was obsolete or not comprehensive.

DOI: 10.4018/978-1-61350-311-9.ch006

The void of decisive leadership by top management allowed the conflicts between different entities in the organization to go on in an increasing mode until the end of the project, which had a negative effect on the project success. The new system design was not successful in resolving the ownership of the data within the organization. This was an issue that caused user resistance to the project.

INTRODUCTION

This case study revolves around a governmental public service institution, which receives public and government money which it invests in. The organization is subject to the supervision of a government Minister who chairs its Board of Directors which also consists of a representative from three ministries, and the national chamber of commerce, the trade union federation and three expert members. The assets owned and managed by the organization are equal to about US$10 billion and yearly revenue equals about US$1 billion. The organizational structure of the Institution consists of five main sectors; General Administration Sector, Servwork Sector, Automation Sector, Investment Sector, and Public Service Sector.

The organization was among the first in the country to be computerized and the first to depend heavily on IS/IT in conducting its work processes. This was a strategic decision made by the founding chairman in 1979. Not long after the formation of the organization, the old/current system was built, in the early 1980s, by an international software development vendor that trained the newly hired staff who had no previous computer skills or knowledge. The training was system-specific that enabled the staff to take-over the day-to-day operation and maintenance of the system. The system was mainframe-based using COBOL and VSAM and sequential file types for batch processing. Around the mid-1980s a large 'stand alone' personnel information system was developed, based on an inverted-relational database management system, to accommodate about 800 employees. The users of all the organizational information systems are about 90% of the organization's staff (more than 700 employees). In the mid-1980's there was a complete change of the top management, where a different team replaced the founding chairman and his team. This new team remains in charge of the organization until the current time.

Sequence of Events

In 1992, the organization planned to open a branch to provide services to the public in another location in the country and relocate the training department and image file-storage on remote sites. At that time, the organization was facing the problem that the computer master-file had reached the maximum record size allowed by

XXX-VSAM type file, which the organization uses. This is a limitation of such a file type that could be expanded to a maximum limit. If this limit is reached then the solution is to go for a database environment. For those reasons the top management decided to conduct a total re-systemization of the mainly batch COBOL-based systems' environment to a relational database-based system environment and to a new network that can support the proposed branch and new remote sites. The re-systemization decision was made because top management had been informed by external sources that many local organizations have converted to database environment and they were successful. The objectives of the project were to install a new network that could support the branching plan and the remote sites. Also, to conduct a total re-systemization of all the organizational systems to convert from a batch-COBOL-VSAM-based environment into a relational database management systems (R-DBMS)-based environment using a fourth generation language as a host language for coding the application programs. This re-systemization was to include the independent 'stand-alone' personnel system, which is based on an inverted R-DBMS package that uses a query language for data query and manipulation. The project would involve converting all the application programs that make up the organizational systems. A steering committee was formed of the Chairman's Deputy, IT/IS Unit Head, Head of Systems Department, and Head of Operations Department. Later, when a vendor was chosen and the project started, the vendor's team head joined the committee. At the request of top management, the IS/IT unit (Automation Sector) decided to do a feasibility study where they examined the products of three international vendors. They limited themselves to those vendors that could provide the organization with a 'total solution', which meant changing all systems (the system) in the organization to the relational database environment. It also meant that the vendor should have the capability to reprogram/convert all applications to the new system environment. Computer Associates (CA), XXX, and YYY were found to fit this pre-requisite condition.

- CA had a small operation in the country and moved its head office to another country in the region.
- XXX had the R-DBMS "DB2", but their local branch did not have the capabilities for converting the applications. In order to do so, it would subcontract the project to many other companies and act only as a coordinator.
- YYY had a small office in the country, while it had its regional head office in another regional country.

In 1993, after the feasibility study was concluded, the organization formed a team consisting of the IS/IT unit manager and a full-time consultant that was hired to serve in this team. The team was formed for the following reasons:

- To produce a general system specification based on the 'total solution' idea and a report for proposal (RFP). This needed to be done while having in mind the at-most utilization of the existing organizational IT infrastructure, especially the hardware.
- To recommend a vendor to conduct the detailed requirement specifications which was termed 'strategic study'.
- To oversee the detailed requirement specifications of the 'strategic study'
- To recommend a vendor out of the three specified by the feasibility study to develop and implement the system
- To form the team to represent the organization side in the project development team.

After several meetings with the vendor's representatives, the initial recommendation of the team was made in favor of YYY. The final selection decision needed to be made after conducting the 'strategic study' to ensure that the vendor had the capability to deliver the specified systems. The team had requested YYY to conduct the 'strategic study'. This was an independent study from the project contract for the 'total solution'. YYY was chosen because its price to conduct this study was half the price of the lowest bid made by the other companies. Also, the initial study favored YYY; if this study was to be made by YYY, and it was the most likely to be chosen, then some of the output of the study (data entity relational diagram) could be directly incorporated in the CASE tool method which would save the organization time and money, which they would have incurred had they chosen any of the other vendors. In fact, an informal decision was made to use YYY, but for formality reasons, which included government regulations, and so this process had to be followed. The 'strategic study' was conducted by four YYY employees and spanned over four months. The study included interviews with all users' representatives, especially the managers and key persons. Questionnaires were also used in addition to reviewing the existing documents. This study's outcome was to provide the organization with such information as:

- The resources required for the project: staff, time, hardware, etc.
- Document/define all work processes; such documentation/definitions did not exist.
- Determine the interaction between different systems.
- Information regarding data-files' condition, formats, storage media types, size, backups, etc.
- Find opportunities to improve on processes and structure.
- Match the project's requirements to the capabilities of potential products and vendors.
- A data entity relation diagram.

At the end of the study, the organization officially announced YYY to provide the R-DBMS and develop the applications needed in SQL fourth generation language and signed the contract, which was to take two years to fully implement the 'total solution'. The vendor provided the organization with the team members' names and résumés and these people were interviewed and approved. Approval of vendor staff by the organization was not stated as a term in the contract, but was agreed upon by the two parties verbally. The two project teams from both the vendor and the organization worked together in a partnership to provide the joint project team but each had their own leader. The IS/IT manager led the team from the organization side. Because the requirements were defined in the 'strategic study', the joint project team started the work on the design of the system. For the first six months, activities went according to plan. After that, and while still in the design phase, the vendor started to change its team members, reallocating the experienced ones to other projects abroad and replacing them with staff members with less or little experience. The reallocation and replacement of the vendor staff continued throughout the project life. The organizational side of things did not look much better, and departments, both IS/IT and users, demanded that the key staff assigned to the project in the early stages return to their original posts. They claimed that the day-to-day work was negatively affected by their full-time assignment to the project. Top management caved in, and most of these who left the team were replaced with junior, less experienced members. In spite of the low level of experience and skill of those junior staff, no training was provided and they were left to learn the basic skills on the job. As the project progressed and the new system output started to appear, user resentment and criticism started to grow. No effort was made by the project team to clear possible misunderstandings or increase awareness regarding the project's aims, objectives and implications. However, top management support became less. This had its negative effect on the needed approval for changes and modifications in system design, organizational structure, and work processes that would come up during the development phase. User rejections and reservations gradually gained top management sympathy and support. By the end of the project duration, users rejected a large number of the systems' functionalities on the grounds that they did not fulfill the needs for conducting the work tasks of the organization. On both sides, the vendor and the organization, agreed on an extension of the contract with amended penalty terms put by the organization. This was followed by a second extension with no acceptable results. After four years from its beginning, and at a cost of between US$7 to US$10 million, the project was declared a failure and the matter is now in the hands of the courts. One decade has passed, since the initial study for a new system was conducted, and the organization still has its 20 year-old system. At the time of preparation of this chapter, criticism of inefficiency of the organization and waste of money is a hot subject in the country's national media and parliament.

ANALYSIS AND DISCUSSION

The readiness of the organization for an IS/IT project can be depicted by the use of a model that explains particular requirements in terms of four domains embracing eight attributes: IT (systems), people (staff, skill, head of IS/IT function), process (practices), and environment (management style/leadership involvement, structure, and culture). The environment attributes could be classified also as being of the people issues as culture and management style/leadership involvement, or closely tied to its systems as in the case of structure. Each of the attributes is described on six levels/stages where each represents a maturity level describing the organizational situation in terms of the particular attribute. Every element describes an attribute comprising of an aspect of how the status of that particular attribute should be at different organizational IS/IT maturity levels. The levels described by the model do not intend to make a judgmental statement of the status of the organizational maturity. Some of the descriptions of the attributes of the early levels might be understood to have a negative notion. The model is used in specifying the levels of maturity of the different attributes mandated by the pre-selected IS/IT. Those levels are considered to be the needed/target levels for the organization to achieve in order to implement successful IS/IT which should lead to a successful IS/IT project. The difference between the current organizational situation and the needed/target in terms of all the domains' attributes constitutes the readiness gap.

PEOPLE: STAFF

After withdrawal of the experienced staff, the vendor provided inexperienced staff to the project. There were not enough qualified members of staff in the organization who could handle both the on-going daily organizational work and the re-systemization project. Both of those tasks needed total commitment. The organization decided to remove its key people from the project team and return them to their original groups to do their daily tasks, while assigning staff with little or no experience to the project. The junior staff assigned to the project team had no training prior to, or during the project. Due to the complexity of the project and the need for a holistic view of the integrations between the many different sub-systems, training the inexperienced staff assigned to the project was highly unattainable. This is because the current systems were not sufficiently documented, and this makes the knowledge and the know-how reside only in heads of key-people. This is complicated by the nature of the culture that exists in the organization being an individual un-cooperative culture. Staff try to keep their knowledge and experience to themselves for different reasons, including job security. Key people consider that keeping the knowledge to themselves would make them indispensable and increase their value to the organization. Two years from the time the project started, the project leader on the vendor side was changed and the new one had a different approach to the project development; he attempted to change the new system design.

Status Prior to Project

At the beginning of the project, the status mostly matched the general description of level 3. This is because the organization had programmers, analysts, and database administrators for the personnel system, and almost all needed technical staff. Also, the IS/IT staff are coordinated with current and future IS/IT needs at both the organizational and unit levels and new user recruits are expected to have specific IS/IT-related skill. Additionally, the technically-oriented IS/IT manager still has a middle management title. The description of this level in the readiness model is as follows:

- Added to the programmers and analysts, dedicated IS/IT planners and database administrators are appointed
- Almost all required technical specialist staff are in-house
- A technically-oriented IS/IT manager is appointed or DP manager might have a change in title
- IS/IT workforce are coordinated with current and future IS/IT needs at both the organizational and unit levels
- New user recruits are expected to have specific IS/IT-related skill

Status at Termination of Project

The status of the staff at the termination of the project deteriorated to mostly match with the general descriptions of some of level 1 and level 2 characteristics. This is because project team members have changed from both the organization and vendor sides to have less experienced, mostly junior staff. This had an impact on the overall staff status. Level 1 and 2 descriptions are as follows:

Level 1

- No dedicated IS/IT staff or small number of low-level technicians and programmers
- No manager allocated responsibility for IS/IT
- External contractors may be used to develop/install systems as required
- Users' new recruits are not expected to have IS/IT-related skill

Level 2

- The small IS/IT staff consists, in addition to programmers and low level technicians, of system analysts where qualified individuals (mainly programmers and analysts) are selected, recruited, and transitioned into assignments
- DP manager who was recently appointed is responsible for IS/IT function
- IS/IT staff are now charged with the responsibility of adequately understanding the user requirements needed for systems' development
- New user recruits are expected to have basic IS/IT skill

Target Status

The proposed target situation needs to agree with the general description of level 5 because the system spans all organizational functions which needs an IS/IT head that has at least the power of user managers. Also, an organization that depends to a great extent on IS/IT and has been in such a situation for 20 years, needs to have a strategic vision of IS/IT and staff who can plan accordingly. This also requires staff on both the user and IS/IT sides who would understand each other's work to be able to have integration of groups and systems. The organization needs to have core hybrid staff to be developed and retained, and to combine the roles of IS/IT and business planners to plan the strategic IS/IT for individual groups and to the organization as a whole, where the business/IS/IT planners acquire experience from working in/with both users and the IS/IT unit which makes them cross-disciplinary. The organization also needs to restore the IS/IT manager to the senior management status he had a decade ago. The description of level 5 is as follows:

- Core hybrid staff sought, developed, and retained, while in some large organizations, some developed expertise is outsourced
- Combining the roles of IS/IT and business planners to plan the strategic IS/IT for individual groups and to the organization as a whole, where the business/IS/IT planners have experience from working in/with both users and the IS/IT unit which makes them cross-disciplinary
- IS/IT manager has senior management status
- The emergence of innovator workforce in the organization.

PEOPLE: SKILL

The contract signed with YYY included up to US$70,000 worth of training (courses/programmes), and the contract also included technology/experience transfer by having organizational staff work alongside the vendor. The project leader on the organization side noted that he did not have the required skills for such a major project. He did not receive proper training, and had no previous experience. The organization assigned inexperienced staff to the project, while the project needed the best skill and experience. This was true with both the IS/IT unit and user groups. The vendor's experienced staff were withdrawn after six months. Many of the vendor's new staff who came into the project had little experience. In many instances, especially in the advanced stages of the project, the organization's staff were more experienced than the vendor's staff who were supposed to train them. On a few occasions, the organization requested the removal of some of the vendor's staff because it thought they were incompetent. The skills of the organization's IS/IT staff on the project team were mainly of COBOL programming with limited systems analysis. Very limited numbers had the skills to use the inverted relational DBMS and its query language. Training was largely not available for the inexperienced staff who worked on the project. The organization, at some time during the project, provided the staff with video-based lectures to be viewed in the employees' own time. This did not result in considerable success. The vendor did not suggest any training or pre-project skill requirements for the project team members on the organization side before the start of the project.

Status Prior to Project

The status mostly matched the general description of level 2 because users in the organization had the required training and skills to use the IS/IT. The technical expertise was limited in scope where IS/IT project development was mainly done to accommodate new changes in governmental laws and regulations that reflect the way the organization conducts its work. This made project management skills limited to such small projects, but required the organization to provide IS/IT staff who had the relevant training and development opportunities needed to perform such assignments. Since the IS/IT job market has a high turnover, the organization felt obliged to provide its IS/IT staff with remuneration and benefits based on their contribution and value to the organization in order to retain them. The description of this level (level 2) in the readiness model is as follows:

- Users begin to have the required training and skills to use the new IS/IT being developed/purchased
- Still little in-house technical expertise in IS/IT development (methodology, structured techniques) and other important skills
- IS/IT staff acquire the skills needed to develop and maintain complete systems such as programming and analysis, in addition to being able to install off-the-shelf, ready-made packages

- Limited project management skills
- IS/IT individuals have the skills required to perform their assignments and begin to have the relevant training and development opportunities
- Individuals in the organizational workforce have remuneration and benefits based on their contribution and value to the organization.

Status at Termination of Project

The skill status deteriorated at the project termination to mostly match the general descriptions of some of level 1. Project team members have changed from both the organization and vendor sides to have less experienced, mostly junior staff. Level 1 description is as follows:

- Users find it hard to acquire the skills to use the IS/IT that exist and skills are individually based and jealously guarded from others
- The emphasis is on technology rather than organizational or informational issues where there are limited technical skills in the organization/unit
- IS/IT skills are specific to individual IS/IT applications
- Required skills are of low level technical nature and there may exist limited advanced (programming or systems analysis) skills in the organization
- There is almost no IS/IT training provided by the organization.

Target Status

The proposed target situation needs to comply, at least, with the general description of level 4, in addition to the strategic planning capabilities and senior management skills in level 5. This is because the system spans all organizational functions, which needs an IS/IT head who has the senior management skills. Also, an organization that depends to a great extent on IS/IT needs to have a strategic view of IS/IT, and staff that can plan accordingly. This also requires good communication skills to enable the integration of the different organizational groups. The descriptions of levels 4 and 5 are as follows:

Level 4

- Systems staff have business skills, because business knowledge and skills are required now of IS/IT staff, besides technical capability, to fit in more with the rest of the organization; at the same time, users gain proper insight into IS/IT related issues
- IS/IT Head and staff have good interpersonal skills

- All individuals are involved in capturing/documenting their knowledge and experience from performing IS/IT-related work to be used in enhancing their competency and performance
- Organization's workforce have the ability to mentor, that is to use the IS/IT- related experience to provide personal support and guidance and to share professional and personal skills and experiences with less experienced staff, with the goal of development of these individuals. This guidance can involve developing knowledge, skills, and process abilities, improving performance, handling difficult situations, and making career decisions.

Level 5

- Core technical skills are developed, and some expertise might be outsourced
- IS/business planners have the skill and experience to plan the strategic information systems for individual units and the organization as a whole, where they gained this experience from working in/with both users and IS/IT function
- As the IS/IT function becomes an integral part of the organization, hybrid skills are used wherever possible, and entrepreneurial skills start to be encouraged within the IS/IT and user workforce, while very knowledgeable users of IS/IT become quite normal, where they now contribute freely to the whole IS/IT operation without any sensitivity from IS/IT function personnel
- IS/IT Head has senior executive skills
- Individuals and workgroups are continuously improving their IS/IT-related capability for performing their work processes.

PEOPLE: HEAD OF IS/IT

IS/IT unit head and the project leader had a middle management status and the head of the project did not have enough authority to implement changes and modifications needed mainly in the organizational structure and processes.

Status Prior to Project

At the beginning of the project, the situation mostly matched the general description of level 4, where the IS/IT manager has middle manager status.

Status at Termination of Project

The situation of the head of IS/IT at the termination of the project remained the same as it was at the beginning, at level 4.

Target Status

The proposed target situation needs to match the general description of level 5, so that the IS/IT head acquires the power needed to manage and enforce policies of IS/IT that span organization-wide, as well as the ability to participate in strategic decisions. There is an IS/IT manager who has senior manager status.

ENVIRONMENT: LEADERSHIP

The leadership support to the project was strong in the beginning, but as time went on, the support became weaker. The prolonging of the project affected top management support negatively. Top management attitude towards IS/IT fluctuated according to the external and internal pressures. For example, if the media focus on a certain aspect of IS/IT in the organization then that aspect receives priority over others. Also, internally-spread rumors could affect the amount of support given to a certain IS/IT aspect or in general. Top management gave priority to the daily operation of the organization over the project, and they relied on acquiring their IS/IT related knowledge from external sources and largely on what other organizations were doing. Even though the organization was heavily dependent on IS/IT in its day-to-day operations, top management considered IS/IT as a tool necessary for smooth functioning, while the heavy dependency on IS/IT should have called for viewing IS/IT with strategic consideration and its flexibility as an asset. At the time when no system output had yet been produced, top management support was high. As the system development progressed, users started to view the output, and top management started to receive many complaints and rumors, which caused the support and enthusiasm to lessen. The incentive approach in the organization is individual-based not team-based, which did not support knowledge sharing and cooperation.

Status Prior to the Project

At the beginning of the project, the status mostly matched the general description of levels 2 and 3 because the attitude of top management fluctuated according to the external and internal pressures. It is important to note that the cost issue was of little importance for top management in the organization because it was a governmental institute that had, at hand, a large amount of income from government and other institutions and a large amount of return on its large investments. Level 3 leadership in a government-run organization would consider IS/IT mainly as a utility to provide services. The descriptions of these two levels in the readiness model are as follows:

Level 2

• Considers IS/IT to be the concern of technologists not management but supportive of IS/IT, where priority and thrust are to minimize the expense of IS/IT utilization.

Level 3

• Considers IS/IT to be one of the many ways to cut costs in the firm and the expenditure on IS/IT as a way of saving cost, where IS/IT is considered as a utility that provides service at minimum cost.

Status at Termination of Project

The status of the leadership at the termination of the project mostly matched with level 2 descriptions. Top management lost enthusiasm and left the project team with little support.

Target Status

The proposed target situation mostly matches the general description of level 5's strategic view of IS/IT, because the organization is an important national service provider that depends almost entirely on IS/IT in running its work processes. In the case of lack of success of the organizational IS/IT, a national problem might occur. As this chapter is being written, such a problem is formulating and complaints regarding the inefficiency of the organization's management are widespread in the media and in the parliament. The description of leadership's level 5 - considers IS/IT as one of the vital parts of the competitive strategy, where a flexible IT infrastructure is perceived as an asset for competitive edge and is brought up in this way during project justification.

ENVIRONMENT: CULTURE

There was a lack of positive relationships between different groups/ departments in the organization. Different managers were suspicious of the project by thinking it was a plan being executed to deprive them of their powers in favor of others. This caused them to resent the project and not to cooperate with the project team. Key staff considered keeping knowledge and experience to themselves as a job security tool, they were not forthcoming in cooperating with the project team, and the lack of decisive leadership by top management allowed the conflicts between different entities in the organization to go in an increasing mode until the end of the project. Also, security and confidentiality issues were obstacles in taking the incremental approach in building the systems, which was preferred by the vendor. The project team was prevented from incremental changes. Top management did not want them to 'touch' the investment working sub-system when the team saw fit, even though it was covered by the contract to convert it to the database environment. In addition, there was a lack of strong support from the top management for necessary changes in the processes and structure. The vendor did not suggest any changes in the culture, and in fact, the issue of culture was never mentioned. Groups/departments lacked a cooperative work environment. Despite the fact that work processes cut through departments' boundaries, there was no cooperation between them. Also, within each group, the individualistic attitude kept people from sharing knowledge. Some of the key users felt threatened by the new system because it would automate some of their tasks that they believed would give them advantage if they remained un-computerized. The project team did not make an effort to relieve the tension in the relationship that occurred in the organization because of the introduction of the project – and at some point in the project, adopted the user-centric approach in testing the system's functions. The key users at that time became a burden on the project by being negative and demanding all the time.

Status Prior to Project

At the beginning of the project, the status mostly matched the general description of level 1 because the relationship between users and the technical staff was one of support and maintenance for the existing IS/IT, where there was no recognition in the organization of the importance of working towards building a constructive relationship between the IS/IT unit and users or among staff members. The description of level 1 in the readiness model is as follows:

- The relationship between users and the technical staff that may exist in the organization or contracted from outside is of support for the existing IS/IT products.
- There is no recognition in the organization of the importance of working towards building a constructive relationship between the IS/IT function and users or among staff members at the same group.

Status at Termination of Project

The status of the culture at the termination of the project remained the same as it was at the beginning, on level 1.

Target Status

The proposed target situation should, at least, match the general description of level 4 and preferably level 5, because the system is a highly integrated system that cuts through unit and group boundaries. In such an environment, the relationships between all organizational workforces need to be those of integration and coopera- tion. The organization needs the IS/IT unit to support the activities of users, and both cooperate on equal bases as partners. There also needs to exist, an emphasis on organizational integration, between different workgroup capabilities and knowl- edge on both IS/IT unit and user sides. Those workgroups (IS/IT and users) need to have the responsibility and authority for determining how to conduct their business activities most effectively. The descriptions of these two culture levels, 4 and 5 in the readiness model, are as follows:

Level 4

- IS/IT function supports the activities of users
- There exists an emphasis on organizational integration between workgroups among which is the IS/IT function.
- Workgroups (IS/IT and users) have the responsibility and authority for deter- mining how to conduct their business activities most effectively
- An improvement in the efficiency and quality of interdependent work, re- sulted from the integration of the capabilities and knowledge of different workgroups on both IS/IT function and user sides, and with each other.

Level 5

- IS/IT function and users cooperate on equal bases as partners
- In addition to the existence of the characteristics stated in the second, third, and fourth points of level 4 above, there exists a continuous striving for inte- gration of organizational workgroups.

ENVIRONMENT: STRUCTURE

Although the IS/IT unit in the organizational chart has a senior status through the 'technical office' of the 'Automation Sector', the new top management, unlike the previous one, left the position empty for over a decade. The head of the IS/IT unit is the same as the project leader, who has a middle management position. Structure changes were obviously needed as implied by the IS/IT unit head, the vendor did not suggest any changes in the 'strategic study'.

Status Prior to Project

At the beginning of the project, the status mostly matched the general description of level 3, because the IS/IT unit technically-oriented manager has a middle management authority and status, where management of the IS/IT unit is centralized and seeks control of IS/IT matters in the organization. Also, there are no organizational IS/IT architecture policies and standards; rather, there are accepted lists for choices of preferred suppliers of IS/IT products. The description of this level in the model is as follows:

- Official power is vested in the IS/IT function, where a new technical IS/IT manager is appointed or the DP manager might have a change in title to IS/IT manager which goes with a similar change in department name
- There are organization-wide IS/IT architecture policies and standards for telecommunications, preferred suppliers, e-mail, etc.
- Management of the IS/IT function is centralized
- IS/IT staff seek control of IS/IT matters.

Status at Termination of the Project

The status of the structure at the termination of the project remained the same as it was at the beginning, at level 3.

Target Status

The proposed target situation needed to match the general description of level 4 as for the need to have architecture policy and standards for coordination, implementation and utility of IS/IT, but also to emphasize (as in level 5) that the IS/IT unit and project team need to have senior authority, to execute their mission to implement a strategic coalition and partnership with user groups based on a strategic view of IS/IT. The descriptions of levels 4 and 5 are as follows:

Level 4

- IS/IT function is well established and its mission is to exploit the IS/IT for business purposes and provide competitive IS/IT in a partnership environment with users
- Decentralized responsibility of IS/IT services with central standards and policy for coordination, implementation and utility
- Units' IS/IT function reports to units' business manager
- Significant degree of involvement of users in IS/IT-related decisions, where IS/IT investments are derived from users' stated needs.

Level 5

- Federal decentralized management structure with flexibility to support IS/IT initiatives, where there exists a strategic coalition and partnership between the IS/IT function and user groups in large organizations
- Decentralized IS/IT function units, with a central IS/IT function providing an organization-wide communication system, major data processing, and large-scale hardware in large organizations
- In some organizations, the budget of the central IS/IT function is paid for by business units for services rendered.

IT: SYSTEMS

There were problems with the format of the data-files in the organization. The size of the master file record size reached the maximum that is allowed by VSAM file format. The data that was coming from external governmental entities was written in a non-compatible format that required special, time-consuming, conversions. Also, this governmental data was not reliable, with bad data both in context and in content that required editing procedures, both computerized and by people. The new system design was not successful in resolving the ownership of the data within the organization. The vendor treated the project as if it was building a new system in an organization that had no prior systems. The new design changed all the interfaces and environment that the user was used to, which caused resentment by the user. After two years in the project, the vendor changed the system design, which caused unease for project members on the organization side.

Status Prior to Project

At the beginning of the project, the status mostly matched the general description of level 4 but in an unfulfilled way. This was because all required operational IS/IT was mostly in place but did not fully serve the organizational needs, and office automation existed, but in an isolated, not standardized manner. Also, there existed an organization-wide network, where all groups were connected and the central IS/IT unit provided communication services for all groups, but it did not accommodate the planned expansion and did not support e-mail. The description of level 4 in the readiness model is as follows:

- All required operational IS/IT is mostly in place and some DSS start to appear
- Office automation is integrated and unified/standardized organization-wide
- Existence of an organization-wide network, where all groups are connected and the central IS/IT function provides communication services for all groups in the organization
- Central coordination in the use of IS/IT throughout the organization, where an effort is made by groups' IS/IT functions to follow standards set centrally
- The organization-wide network is starting to be utilized to connect users to whatever shared applications and information systems that are needed
- Extensive use of standard e-mail messages throughout the organization, and there is evidence of dependence on the organization-wide network to conduct formal communication.

Status at Termination of the Project

The status of the systems at the termination of the project remained the same as it was at the beginning, at weak level 4.

Target Status

The proposed target situation still needs to match the general description of level 4 but with a new network infrastructure that utilizes e-mail and connects to the Internet to facilitate a better service for the external customers.

PROCESS: PRACTICES

From the start, processes were not defined and documented prior to the project; the vendor suggested top management and users allowed no process change when the need was recognized by the project team later in the project. Due to the inability to reengineer processes, 'corner-cutting' and 'go around' techniques were used, such as allowing two units to access, modify, and delete the same data. This caused problems and conflicts for the project team with users, and among user groups.

Status Prior to Project

At the beginning of the project, the status mostly matched the general description of level 2-GG 1 - Identify Work Scope and Perform Base Practices.

Status at Termination of Project

The status of the process at the termination of the project improved to include some of the generic practices of level GG 2. Those additional practices were the result of the 'strategic study' conducted after the start of the project and are designated GP 2.1, GP 2.2, and GP 2.7. The level description is as follows:

- GP 2.1 Establish an Organizational Policy
- GP 2.2 Plan the Process
  1. Obtain management sponsorship for performing the process.
  2. Define and document the process description.
  3. Define and document the plan for performing the process.
  4. Review the plan with relevant stakeholders and get their agreement.
  5. Revise the plan as necessary.
- GP 2.3 Provide Resources
- GP 2.4 Assign Responsibility
  1. Assign overall responsibility and authority for performing the process.
  2. Assign responsibility for performing the specific tasks of the process.
  3. Confirm that the people assigned to the responsibilities and authorities understand and accept them.
- GP 2.5 Train People
- GP 2.6 Manage Configurations
- GP 2.7 Identify and Involve Relevant Stakeholders
  1. Identify stakeholders relevant to this process and decide what type of involvement should be practiced.
  2. Share these identifications with project planners or other planners as appropriate.
  3. Get stakeholders involved as planned.
- GP 2.8 Monitor and Control the Process
  1. Measure actual performance against the plan.

2. Review accomplishments and results of the process against the plan.
3. Review activities, status, and results of the process with the immediate level of management responsible for the process and identify issues.
4. Identify and evaluate the effects of significant deviations from the plan.
5. Identify problems in the process and in the plan.
6. Take corrective action when requirements and objectives are not being satisfied, when issues are identified, or when progress differs significantly from the plan.
7. Track corrective action to closure.
*   GP 2.9 Objectively Evaluate Adherence
*   GP 2.10 Review Status with Higher-Level Management

Target Status

According to the project manager, the proposed target situation needs to match the general description of level 4-GG 3. The description of level 4-GG 3 is as follows:

*   GP 3.1 Establish a Defined Process
    1. Select the standard process that best fits the specific instances from the organization's set of standard processes.
    2. Establish the defined process by tailoring the selected standard processes and other process assets according to the organization's tailoring guidelines.
    3. Ensure that the organization's process objectives are appropriately addressed in the defined process.
    4. Document the defined process and the records of the tailoring.
    5. Revise the description of the defined process as necessary.
*   GP 3.2 Collect Improvement Information
    1. Store process and product measures in the organizational measurement repository.
    2. Submit documentation for inclusion in the organizational library of process-related assets.
    3. Document lessons learned from the process for inclusion in the organizational library of process-related assets.
    4. Propose improvements to the organization's process assets.

SUMMARY AND FINDINGS

Organizations need, in the contract, to have the qualifications of the vendor's staff, and agree that prior approval for any change of staff or new recruitment would be agreed beforehand. This is because the vendor's staff had a high turnover. Experienced staff which were agreed upon by the organization were assigned to the project for a short time, only at the beginning of the project. Many of the vendor's new staff who came into the project had little experience, and many said that on occasion they were "filling time between jobs" until they found a better job in their own countries, plus, the vendor used the project to train their own, new, inexperienced staff. It is important to recognize when measuring the organizational readiness to focus on the staff in the organization that will be available for the project. There might be competent staff in the organization, but who will not be available. The vendor asked for a key person of each of the users' functions to participate in the system testing to approve the new systems' functions. This made the approval to be subject to this one person's prejudices and reservations. It could have been different if the approval was through a team, representing each user's function.

The lack of positive relationships between different groups in the organization caused resistance to the required changes in structure and processes. Because key staff considered keeping knowledge and experience to themselves as a job security tool, they were not forthcoming in cooperating with the project team. This was complicated by the almost complete absence of systems' documentation, and the little documentation that did exist was obsolete or not comprehensive. The individualistic culture almost entirely prohibited the existence of in-house training programmes for new and junior staff, by senior and key staff. These programmes, that were rarely conducted, were largely ineffective because they lacked the practical know-how knowledge and documentation.

The void of decisive leadership by top management allowed the conflicts between different entities in the organization to go on in an increasing mode until the end of the project, which had a negative effect on the project success. Furthermore, the shortage of support from the top management prevented the project team from implementing necessary changes in the processes and structure. The project team did not make an effort to relieve the tension in the relationship that occurred with users that was caused by the introduction of the project. They should have introduced a user awareness programme to explain the benefit of the project and work towards eliminating unfounded suspicions. Level 3 Leadership does not fully apply in this government institution case study. Such a government-run organization would consider IS/IT mainly as a utility to provide services. This does not call for a modification to the model, but requires caution when using the model for government organizations. The project leader thinks that the vendor's main concern at that time was to win the contract, so the vendor stayed away from highly sensitive areas such as culture, processes and structure that might have caused it to lose the bid. As the top management caved in under the pressure of both the IS/IT unit and user groups, the project team was not allowed to modify the organizational structure when it was obviously needed to solve some of the design problems that appeared later in the project.

The new system design was not successful in resolving the ownership of the data within the organization. This was an issue that caused user resistance to the project. The vendor treated the project as if it were building a new system in an organization that had no prior systems. The new design changed all the interfaces and environment that the user was used to. This caused resentment by the user. The new design was understood to imply to the users to forget their experience and knowledge that they had built over many years. Many key users had given their value in the organization and had considerable knowledge and experience, but their perception was that the organization would start afresh if this system was to be implemented. This problem is even more dramatic than it sounds because most of the people, and especially the key staff on both IT/IS and users sides, knew of IT/IS only through the current old system, and at this organization only. They had not been subjected to IS/IT in any other context. Almost all of them had had no IS/IT training either at school or in any other context. Almost all of them come from non-technical educational disciplines and backgrounds. They were subjected to IS/IT and trained only in this organization and on this system. Most of the IS/IT personnel did not even have their own computers at home. Working with IS/IT for many years at one of the pioneering organizations in introducing IS/IT in the country became a personal prestige that produces a social status in addition of being a skill and a career, especially for those in the IS/IT unit who carried the "Systems Engineer" title. To think that the new system would deprive them of all of the status they had gained and return them to square one by becoming trainees, instead of the experts, was a cause of strong resentment by almost all the key organizational staff.

The 'strategic study' did not address the real problems with the system. For example, the user managers had to answer a questionnaire with a multiple-choice answer regarding enhancing the performance indicators of the work task. Many of the performance indicators were not applicable to the actual situation. Those indicators were important for the vendor's own experience in other countries' environment where the questionnaire was originated, such as saving time and money, customer satisfaction, *etc.* Those were the not the actual indicators for the user manager.

The withdrawal of the vendors' experienced staff that had been agreed upon had negatively affected the organization's trust in the vendor. The organization did not have this issue in the contract and took the vendor's initial agreement as being enough of an assurance. Also, the organization understood that the vendor would do its best for the project to succeed because it would give the vendor a good reputation in the local market.

The readiness scoring system for each of the four domains embracing eight attributes: IT (systems), people (staff, skill, head of IS/IT function), process (practices),

Table 1. IS/IT Readiness Scoring System for Case Study

Status	Inception	Termination	Target	Gap
People – Staff	3	1-2	5	3,4,5
People – Skill	2	1	4-5	2,3,4,5
People – Head of IS/IT	4	4	5	5
Environment - Leadership	2-3	2	5	3,4,5
Environment - Culture	1	1	4-5	2,3,4,5
Environment - Structure	3	3	4-5	4,5
IT – Systems	4	4	4	Improvements
Process - Practices	2-GG 1	3-GG 2	4-GG 3	3-GG 2

and environment (management style/leadership involvement, structure, and culture) for this case study is as shown in Table 1.

CASE STUDY ADOPTED FROM

Al Saleh, Y. (2002), "IS/IT Success and Evaluation: A General Practitioner Model", PhD Thesis, School of the Built Environment, The University of Salford.

Section 2
Issues in IT/IS Readiness

Chapter 7

The Ways of Assessing the Security of Organization Information Systems through SWOT Analysis

David Rehak
VSB – Technical University of Ostrava, Czech Republic

Monika Grasseova
University of Defence, Czech Republic

EXECUTIVE SUMMARY

The chapter is focused mainly on assessing the factors of the external environment in the area of security of information systems in the organization through SWOT analysis. At first the method is characterized from the viewpoint of its purpose and nature. The emphasis is laid on the principles of SWOT analysis, the possible use of methods and tools, and also the most common problems occurring during the implementation of the analysis. The recommended methodical procedure for the implementation of SWOT analysis is described in another part of the chapter with individual phases and particular activities, which are appropriate to be carried out within these phases. The main part of the chapter is focused on the ways of semi-quantitative assessment of threats to the area of information systems of the organization, while evaluating their risks, and the assessment of opportunities, while evaluating their benefits.

DOI: 10.4018/978-1-61350-311-9.ch007

Both cases include a detailed description of procedure leading to an objective outcome during the classification of identified threats and opportunities according to the set criteria.

INTRODUCTION

The assessment of identified threats and opportunities is a significant phase of SWOT analysis, which may fundamentally affect security of information systems of the organization. The most objective outcomes have to be achieved in this phase on the basis of which an optimal development of organization will be chosen. The assessment of threats and opportunities while evaluating their risks and benefits is one possibility of achieving such objectivity. An accurate assessment may also be achieved with the use of a multi-criterial assessment matrix. The chapter is aimed at clarifying the ways of assessing the identified threats and opportunities in the external analysis of the security of information systems in the organization within SWOT analysis.

SETTING THE SCENE

SWOT is an acronym for *Strengths, Weaknesses, Opportunities*, and *Threats*. Thus SWOT is the acronym for the internal strengths and weaknesses of organization and the opportunities and threats identified in the external environment of organization. SWOT analysis is one of the methods of strategic analysis of the initial state of an organization and/or its parts, generating the alternatives to strategies (see Figure 1) on the basis of internal analysis (strengths and weaknesses) and external analysis (opportunities and threats). A comprehensive SWOT analysis puts strengths and weaknesses of an organization or its parts against identified opportunities and threats ensuing from the surrounding environment and defines the position of the organization and/or its parts as a starting point for defining the strategies of further development.

The method was developed by Albert Humphrey, who led a research project in the 1960s-1970s at Stanford University. The project was financially supported by the 500 biggest corporations in the USA (Fortune 500) and its aim was to analyze shortcomings in the planning process of those corporations and develop a new system of change management for them. A team method for planning was called SOFT analysis and later revised as SWOT analysis.

Figure 1. The basic framework of SWOT analysis

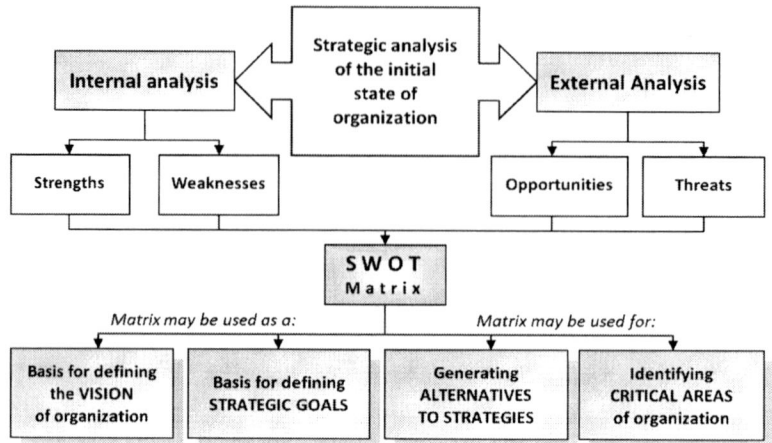

SWOT analysis may be included among the most implemented analytical methods. Specialized literature usually includes only the outcome of the last phase of SWOT analysis, i.e. SWOT matrix (see Figure 2).

During SWOT analysis it is necessary to determine the purpose of its use, i.e. what the outcomes will be used for. SWOT analysis may be used for one or more of the following purposes:

- As a basis for defining the vision
- As a basis for defining the strategic goals

Figure 2. SWOT matrix

Internal Factors / External Factors	Weaknesses (W) 1. ...., 2. ...., 3. ...., etc.	Strengths (S) 1. ...., 2. ...., 3. ...., etc.
Opportunities (O) 1. ...., 2. ...., 3. ...., etc.	**WO strategy** „Searching" *Overcoming a weakness by taking advantage of opportunity*	**SO strategy** „Taking advantage" *Taking advantage of strength in favour of opportunity*
Threats (T) 1. ...., 2. ...., 3. ...., etc.	**WT strategy** „Avoiding" *Minimization of weakness and avoidance of threat*	**ST strategy** „Confrontation" *Taking advantage of strength to prevent threat*

- As a basis for the first generation of strategic alternatives
- For identifying critical areas.

Many organizations finish SWOT analysis with a detailed list of strengths, weaknesses, opportunities and threats. However, if the facts discovered are not used for the purposes as outlined above, the findings are basically useless. The question is, what the purpose of discovering the weaknesses of the organization is, e.g. in securing the information systems, if the organization does not work with such information any more. Many organizations carry out SWOT analysis just to claim it has been completed during the preparation of the information systems security crisis plan, for example. However, the fact that the plan does not reflect the outcomes of analysis is not considered. Therefore when implementing SWOT analysis it is necessary to consider the purpose of it and the further use of outcomes.

The analysis has not had a fixed methodological framework so far. General information on the SWOT analysis procedure is published in specialized literature rather than particular steps accompanying its practical implementation. Therefore one subchapter includes a recommended methodical procedure of implementing SWOT analysis. The procedure cannot be applied as a universal one, but it is necessary to amend it according to particular conditions, specifics of an organization and purpose of analysis.

The following principles are to be followed during SWOT analysis:

- The purpose has to be considered all the time during the analysis; procedures and outcomes cannot be mechanically applied to a different problem;
- It is necessary to focus on substantial facts; the statement of strategy is complicated in case of too much information. SWOT, as part of strategic analysis, should identify only "strategic" facts, i.e. long-term phenomena;
- Analysis has to be objective - it can be achieved with more people participating in its development;
- It is appropriate to use the system of assessing the power of factors, e.g. by using point scales.

The most often used methods and tools of SWOT analysis are as follows:

- The use of data from assessment and analytical reports and studies – it is usually the content analysis of elaborate documents, which include some type of analysis, either the analysis of the initial state or the prognosis of future development;

- The implementation of creative methods (e.g. brainstorming, panel discussion) and procedures based on professional forecasts made by competent entities;
- The implementation of appropriate forms, matrixes, graphs and point scales.

Characteristics of SWOT Analysis

SWOT is a type of strategic analysis of a company or an organization from the viewpoint of its strengths, weaknesses, opportunities, and threats. It provides data for stating the development directions and activities, company strategies and strategic goals. The analysis is based on the analysis and assessment of the current state of the organization (internal environment) and the current state of the surrounding environments of the organization (external environment).

The strengths and weaknesses of the organization are identified in the internal environment. Strengths and weaknesses define internal factors of effectiveness and ineffectiveness in all significant areas of the organization. The organization may be divided into the functional and procedural areas, or it may use a "7S" model (see part Preparation for SWOT analysis, step 2).

Opportunities and threats for an organization are identified in the external environment. Opportunities and threats define the effects of the external environment in all significant areas of the organization. The external environment affecting organizations in public and private sectors usually includes the following areas: P – political, E – economic, S – social, T – technological, L – legislative, E – environmental. The analysis of opportunities and threats may be carried out with the use of PESTLE analysis. The list of opportunities for a public sector organization may include, for example, available financial resources, public interest, interest of certain segments of society resulting in stimuli towards improved quality of services and international co-operation, and analysis of the political environment. The list of threats may include, for example, the outcomes of surveys of competitive environment within a selected segment, limited financial support provided to prepared projects from public resources and supranational organizations, lack of inventions and innovative processes in the area of interest, negative development processes of the national economy (macro and micro-processes of national economy), etc. Although the organization cannot affect external factors as much as internal ones, it should take appropriate measures to minimize them (in case of threats) or take advantage of them (in case of opportunities).

RECOMMENDED METHODICAL PROCEDURE FOR THE IMPLEMENTATION OF SWOT ANALYSIS

It is appropriate to start from the general principles for the implementation of SWOT analysis. There are four basic phases of SWOT analysis resulting from the lessons learned, regardless of whether it is the production sector or public administration: 1. Preparation for SWOT analysis; 2. Identification and assessment of strengths and weaknesses of the organization and/or its areas (SBU); 3. Identification and assessment of opportunities and threats from external environment; 4. Development of SWOT Matrix.

Individual phases of SWOT analysis are further divided into particular activities and steps. The described procedure of implementing each phase of SWOT analysis is based on proven practical experience and is not binding. As the method does not have a fixed methodological framework it is possible to amend the proposed procedure according to the needs and established practices of the organization and the level of strategy for which SWOT analysis is used.

Preparation for SWOT Analysis

Four steps are proposed to be followed in order, as follows: 1. Clear statement of the purpose of SWOT analysis; (2) Definition of the areas to be analyzed; 3. Establishment of analytical teams; 4. Standardization of work methodology and motivation of team members.

1. Clear Statement of the Purpose of SWOT Analysis

The purpose of SWOT analysis has to be stated, in case it has not already been stated by an employee who ordered the analysis.

2. Definition of the Areas to be Analyzed

In case SWOT analysis is applied to all of the organization, it is suitable to divide the organization into areas, which are then analyzed independently. The organization may be divided according to functional areas, procedural areas or according to McKinsey's "7S" framework. The division of the organization into functional areas may have the following structure:

- Management systems
- Organizational structures
- Information systems

- Culture of organization
- Human resources and their development
- Research and development, equipment
- Finance and economy.

The analysis of the following areas is carried out in case the organization is divided according to the process areas:

- Main processes
- Management processes
- Supporting processes.

The Mc Kinsey 7S model of internal analysis may be used in the analysis of organizations. SWOT analysis can then be used with focus on the following areas:

- Strategy
- Structure
- Management system
- Style of management
- Staff
- Skills
- Shared values.

If a strategic business unit (SBU) of the organization is analyzed it is also suitable to divide it into individual areas.

SWOT analysis may also be used for self-assessment of organization performance. The European Foundation for Quality Management (EFQM) Excellence Model is used for such an assessment in Europe. The model of Common Assessment Framework as the amended and simplified version of EFQM is used in the area of public administration. The self-assessment of organization performance is carried out on the basis of 9 criteria and SWOT analysis may be used as a basis for analyzing the individual criteria. Mainly internal analysis is used for identifying strengths and weaknesses, which are called the opportunities for improvement in this model. The Baldrige Criteria for Performance Excellence Framework is used in the USA and is similar to the above mentioned models.

The determined areas are then usually assessed independently with SWOT analysis. The SWOT analysis carried out within the organization security environment may assess independently the following areas: information systems, technological operations, work safety and health protection at work, etc.

3. Establishment of Analytical Teams

The analytical teams of experienced personnel identify and assess the factors affecting the analyzed areas. It is possible to establish a special team for each area, i.e. each area will be analyzed by a different team, which is the most knowledgeable and experienced in the given area.

4. Standardization of Work Methodology and Motivation of Team Members

All team members must agree on a particular procedure of SWOT analysis and adhere to it. It is suitable to determine the possibilities of collecting information and the methods to be used (or which are recommended to be used). The main motivation of the team members is the fact that they know the purpose of SWOT analysis and that their work will not be useless. The team members must have enough time and competencies in case they acquire certain information in discussion with the management of the organization.

Identification and Assessment of Strengths and Weaknesses

We propose two consecutive steps, which are described in more detail below: 1. Identification of strengths and weaknesses; 2. Assessment of strengths and weaknesses.

1. Identification of Strengths and Weaknesses

Strengths and weaknesses of the analyzed area of the organization may be identified in several ways, e.g. through the content analysis of initial data and the following implementation of creative methods, e.g. brainstorming, consultations, and guided discussions aimed at identifying or defining strengths and weaknesses of the analyzed area of the organization. It is necessary to record the strengths and weaknesses appropriately e.g. in a form, including the justification of the outcome. A sample form for the identification of weaknesses is shown in Table 1.

2. Assessment of Strengths and Weaknesses

The identification of strengths and weaknesses of the analyzed area is usually followed by determining their relevance from the viewpoint of their consequences for the analyzed area. The relevance of strengths and weaknesses is assessed separately using the method of pair comparison or the 100 points method.[1]

Table 1. Sample form for the identification of weaknesses

The analyzed area of organization: *e.g. Security of Organization Information System*	
WEAKNESSES	**WHY?** (justification - why we consider a particular factor to be a weakness)
A. Imperfect updating of information system	*Security gaps occur due to imperfect updating of information system. Malware may then infiltrate such information system.*
B. Infiltration of personnel	*It is unlikely that the company management detects personnel infiltrating the organization information system. Therefore the unauthorized use of data in the information system is quite extensive.*
C. Weak information infrastructure	*The data flows are programmed incorrectly in the information system. It can cause unintended and serious data leakage.*
D. Weak communication infrastructure	*The late updating of hardware may result in possible security errors, which enable malware to infiltrate the information system through unprotected ports.*

The procedure of determining relevance or the order of individual strengths and weaknesses through the method of pair comparison is as follows (see Table 2):

- Identified strengths/weaknesses are compared in pairs and their relevance is determined in relation to the analyzed area. The more important element of each pair is recorded in the table.
- The frequency of higher relevance is counted, i.e. the number of preferences in the pair comparison is counted. The values are summed both in lines and columns.
- The relevance – scale – of each strength/weakness is calculated by dividing the number of preferences of a particular strength/weakness with the total number of preferences (e.g. the relevance of A weakness is $A = 2/6 = 0,33$).

Table 2. The determining of relevance with the method of pair comparison of identified strengths and weaknesses

Process of determining the relevance	a)				b)	c)
Strength/Weakness	A.	B.	C.	D.	Number of Preferences	Relevance
A. Imperfect updating of information system		B	A	A	2	0,33
B. Infiltration of personnel			B	B	3	0,5
C. Weak information infrastructure				D	0	0
D. Weak communication infrastructure					1	0,17
Total					6	1,0

If the 100 points method is implemented, each team member divides 100 points among individual strengths. The more points are allocated to the given strength the more relevant it is considered to be in relation to the analyzed area. 100 points are similarly divided among weaknesses. Relevance may then be calculated as an arithmetic average from the individual assessments of analytical team members. If the assessment of any strength/weakness is considerably different among the team members, it is suitable to reach consensus rather than use the arithmetic average.

Once the strengths and weaknesses are identified they are arranged in the order of relevance according to the results of assessment. Thus two lists are made starting from No 1 as the most relevant strength/weakness to number X as the least relevant one with the least or no consequence for the analyzed area. The order of weaknesses is made on the basis of assessment outcomes (Table 2) and according to their importance as follows:

1. Infiltration of personnel
2. Imperfect updating of information system
3. Weak communication infrastructure
4. Weak information infrastructure.

Identification and Assessment of Threats and Opportunities from External Environment

We propose three consecutive steps, which are described in more detail below: 1. Identification of threats and opportunities from external environment; 2. Assessment of threats; 3. Assessment of opportunities. Steps 2 and 3 are described in separate subchapters, thus only basic information is included in this part.

1. Identification of Threats and Opportunities from External Environment

Threats and opportunities from the external environment may be identified in several ways for the analyzed area of the organization. They can be identified through content analysis of initial data followed by some of the creative methods, e.g. brainstorming, consultation, guided discussion aimed at identifying or defining threats and opportunities of the analyzed area of the organization. It is necessary to record the threats and opportunities appropriately e.g. in a form, including the justification of outcomes. A sample form for the identification of threats is shown in Table 3.

Table 3. The sample threat identification form

The analyzed area of organization: *e.g. Security of Organization Information System*	
THREATS	**WHY?** (justification - why we consider a particular factor to be a threat)
A. Monitoring of network	*Weaknesses are detected and data acquired from the information system in order to prepare future attack or compromise the users of information system*
B. Alteration of sent data	*Data are falsified with the aim to introduce disinformation into the information system.*
C. Insertion of disinformation into the information system	*Direct insertion of disinformation (redundant information) into the information system in order to compromise the users of the system.*
D. Overloading of information system	*Make the information system inaccessible by disconnecting it from the communication infrastructure and high-capacity computer networks with the aim to withhold service.*

2. Assessment of Threats

Assessment of threats is primarily aimed at determining the relevance of consequences of threats from the external environment for the analyzed area if they occur. The probability that individual threats occur is discovered. The level of risk that a given threat will affect the analyzed area of organization will then be calculated as a product of threat and relevance of its consequences on an organization and the probability of its occurrence. The higher the level of risk is the bigger strategic significance it has. Then the risks for the analyzed area are arranged according to their levels. The risk of the highest level will be the risk No 1, etc. The way of assessing threats is described in more detail in a special subchapter below.

3. Assessment of Opportunities

It includes determining the attractiveness[2] of opportunity consequence from the external environment on the analyzed area in case it occurs. The probability of occurrence of individual opportunities is determined, too. The benefit of each opportunity may be determined on the basis of the two variables mentioned above. The benefit is determined as a product of attractiveness of opportunity consequence and the probability of its occurrence. The higher the benefit, the bigger strategic significance it has.

Then the benefits for the analyzed area are arranged according to their levels. The benefit of the highest level will be benefit No 1, etc.

The Development of SWOT Matrix

There are two key activities in this phase of developing the SWOT matrix: 1. The recording of factors of strategic significance; 2. The generating of alternative strategies.

1. The Recording of Factors of Strategic Significance

It is the recording of strengths and weaknesses of high relevance as well as the opportunities and threats of high values (i.e. the benefits and risks of high levels), which are of strategic significance. So it is the selection of those factors (strengths, weaknesses, opportunities and threats), which will be used for the generating of alternative strategies.

2. The Generating of Alternative Strategies

The generating of alternative strategies is based on combining strengths and weaknesses (internal factors) with identified threats and opportunities (external factors). The development of four strategies is then a logical continuation of SWOT Matrix (see part b of SWOT Matrix in Figure 2).

SWOT Matrix includes the following four strategies:

- **The WO strategy – the strategy of searching.** These strategies are aimed at overcoming (eliminating) weaknesses by taking advantage of opportunities. These strategies require obtaining additional resources for taking advantage of opportunities.
- **The SO strategy – the strategy of taking advantage.** These strategies take advantage of strengths in favour of opportunities identified in the external environment. This quadrant specifies the desirable condition towards which the organization is heading. It is clear that these strategies are the basis for defining the visions and goals. The difficulty of such a definition and implementation is given by the fact that the S-O combination occurs rarely in real life.
- **The WT strategy – the strategy of avoiding.** It is a defence strategy aimed at the elimination (overcoming) of weaknesses and the avoidance of external threat. It is the "fight for survival" for an organization. If the strategy is used for the development of concepts it is key for maintaining the fundamental functions of the organization necessary for fulfilling its mission.
- **The ST strategy – the strategy of confrontation.** These strategies are possible to be implemented if the organization is strong enough to be confronted with a threat – basically one group of the organization requires that another group of the organization follows the principles of sustainable development.

CASE DESCRIPTION

Attention will be paid to the assessment of the external environment in the following part of the chapter. The first step, the phase of external analysis (the identification and assessment of opportunities and threats), starts with the identification of threats and opportunities (with the help of brainstorming, consultation and guided discussion). Then the threats and opportunities are assessed with the aim to prioritize them according to the relevance and attractiveness of their consequences. The quantified assessment of threats and opportunities may be carried out by number of methods and tools. The assessment of risks and benefits[3] and the matrix of multicriterial assessment may be recommended as the most suitable methodologies.

ASSESSMENT OF THREATS WITH THE USE OF RISK ASSESSMENT

The process of risk assessment[4] may well be used when assessing the already identified threats. The assessed threats represent certain risks, which increase with the relevance of threats. The risks of individual threats results not only from the relevance of their consequences, but also from the probability of their occurrence. Thus the first step of risk assessment is to determine the relevance of the threat from the external environment and its consequences for the analyzed area in case it occurs. The point scale of five basic levels has been set to assess the relevance of threats (see Table 4).

The second step is aimed at describing the probability of a threat occurring. It can be determined in three ways, qualitatively, semi-quantitatively or quantitatively. All three ways of expressing the probability of the threat/opportunity are shown in Table 5.

The quantitative method describes probability mathematically with the help of the following relation (1)

Table 4. The assessment of the relevance of threat consequences

Verbal description of the relevance of threat consequences	No of points
Negligible	1
Little relevant	2
Relevant	3
Highly relevant	4
Unacceptable	5

Table 5. Threat/opportunity occurrence probability assessment

Qualitative expression of probability	Semi-quantitative expression of probability	Quantitative expression of probability [in %]
Almost impossible	1	⟨1;20⟩
Rarely possible	2	⟨21;40⟩
Commonly possible	3	⟨41;60⟩
Highly probable	4	⟨61;80⟩
Almost certain	5	⟨81;100⟩

$$P = U \cdot X^{-1} \tag{1}$$

where:

P represents probability that a threat occurs;

U represents the number of undesirable events caused by the assessed threat;

X represents the number of all undesirable events, which happened during the existence of organization.

In case we prefer the semi-quantitative way of expressing the probability of threat it is necessary to compare the final value with Table 5 and to convert the probability expressed in percent into a point scale.

Example: The way of quantitative calculation of probability is exemplified by describing the probability of attack against the organization information system in 2011. The period of existence of the organization is 10 years, in this case from 2001 to 2010. During this time there were 80 serious security attacks against the information system, which is the number of all events during the existence of organization ($X = 80$). The second necessary information is the number of attacks, which were not detected in time and caused undesirable impacts on the organization information system, i.e. the number of completed undesirable events of assessed threat ($U = 23$). After inserting the values into the relation (1) it is calculated the probability of attack against the organization information system in 2011, which is 28.8% ($P = U \cdot X^{-1} = 23 \cdot 80^{-1} = 0,288 = 28,8\%$). In this case the assumed event is (i.e. the attack against the information system) is rarely probable in 2011 and its value in points is $P = 2$ in the semi quantitative expression of probability (see Table 5).

The following assessment of threats is carried out with the use of risk assessment by inserting the acquired point values into the relation (2)

$$R = C \cdot P \tag{2}$$

where:

    R represents the assumed level of risk of a given threat;

    C represents the relevance of consequences of the assessed threat;

    P represents the probability that the assessed threat occurs.

The final values of individual threats are then compared between each other and used as a basis for the prioritizing of threats. The most relevant threats are considered to be those which have the highest level of risk.

ASSESSMENT OF OPPORTUNITIES WITH THE USE OF BENEFIT EVALUATION

A similar approach may be used in the assessment of opportunities. However, it is necessary first to define the element, which will affect the final order of identified opportunities. Risks defining the most relevant threats for an organization were such elements in case of threats. The assessment of opportunities can be carried out with the use of benefit evaluation, which includes two steps (as in the case of risks). The first step is determining the attractiveness of opportunity consequence from external environment on the analyzed group of processes in case it occurs. The point scale of five basic levels has been set to assess the attractiveness (see Table 6). The second step is aimed at describing the probability that an opportunity occurs. The probability can also be determined in three ways, qualitatively, semi-quantitatively (see Table 5) and quantitatively (see relation 1).

The following assessment of benefits of the analyzed opportunities is carried out by inserting the acquired point values into the relation (3)

$$B = A \cdot P \tag{3}$$

Table 6. The assessment of the attractiveness of opportunity consequences

Verbal description of the attractiveness of opportunity consequences	No of points
Negligible	1
Little relevance	2
Relevant	3
Highly relevant	4
Fundamentally relevant	5

where:

B represents the assumed benefit of a given opportunity;

A represents the attractiveness of consequences of the assessed opportunity;

P represents the probability that the assessed opportunity occurs.

The final values of individual opportunities are then compared between each other. The opportunities with the highest values are given the highest priority.

ASSESSMENT OF THREATS (OPPORTUNITIES) WITH THE USE OF MULTICRITERIAL MATRIX

The use of multicriterial matrix is another possible and suitable way of assessing the identified threats and opportunities ensuring the objectivity of outcome. In comparison with the previous way of assessment this process can implement, besides the relevance of consequences (or attractiveness) and the level of probability, also other criteria. The criteria will differ depending on the area, which is assessed with SWOT analysis.

The exposure to threat (opportunity) can be considered to be a significant extra criterion. This 'time criterion' introduces another significant factor into the decision-making process, which may fundamentally affect attitude towards the solution of a given area. Another criterion may also be "demand for resources during the development of countermeasures", but its mathematical quantification would probably be demanding and inaccurate, and so its implementation is not recommended.

The first step is based on determining the relevance of consequences (attractiveness) of each threat (opportunity) for a given area, the level of which will be determined with the help of the point scale shown in Table 4 (Table 6 in case of assessing the opportunities). Similarly the probability of each threat (opportunity) will be determined according to Table 5. The exposure (E) will be the last criterion used. It is the ratio of time for which the assessed threat (opportunity) is effective to the time for which the analysis is carried out. This criterion will again be expressed in points from 1 to 5 and its value will be calculated with the help of the following relation (4)

$$E = D \cdot T^{-1} \cdot 100 \tag{4}$$

where:

E represents the final exposure to threat (or opportunity);

D represents the duration of threat (or opportunity);

T represents the time for which the analysis is carried out.

Table 7. The assessment of exposure to threat (opportunity)

Assumed exposure to threat (opportunity) [%]	No of points
⟨1;20⟩	1
⟨21;40⟩	2
⟨41;60⟩	3
⟨61;80⟩	4
⟨81;100⟩	5

The final value expressing the exposure to threat (opportunity) is compared with Table 7 and obtains the corresponding number of points.

Example: Determining the exposure to threat, i.e. the ratio of time for which the threat is effective to the time for which the analysis is carried out, is shown on the following example. The threat "insertion of disinformation into the information system" has been identified within the analysis of external factors of organization. The duration of this threat (D) to the assessed area is estimated to be 3 years, because a new security element protecting the information system against the above mentioned threat is to be installed within 3 years. The time (T) for which the analysis will be carried out is 10 years (i.e. from 2011 to 2020). It can be stated on the basis of the above mentioned that the assumed exposure to this threat is 30% ($E = D \cdot T^{-1} = 3 \cdot 10^{-1} = 0,3 = 30\%$) out of the total time during which the analysis is carried out. The value in points will be $E = 2$.

After determining all criteria, the acquired values are inserted into the prepared Matrix (see Table 9) and the final value of risk (benefit) for individual threats (opportunities) is calculated through arithmetic average (see relation 5). The threats (opportunities) with the highest values are then considered to be the most relevant (attractive).

$$R = C + P + E / 3 \tag{5}$$

where:
R represents the assumed level of risk of a given threat
C represents the relevance of consequence of the assessed threat
P represents the probability that the assessed threat occurs
E represents the exposure to the assessed threat.

Example: The assessment of threats, with the use of multicriterial assessment, is shown in the following case study related to the security assessment of information system in the organization until 2020. The following threats are identified for the above mentioned area within the analysis of external factors:

1.   Threat A (T_A): Monitoring of network
2.   Threat B (T_B): Alteration of sent data
3.   Threat C (T_C): Insertion of disinformation into the information system
4.   Threat D (T_D): Overloading of information system.

Three relevant factors were assessed in the analysis of threats: the relevance of consequences (C), the probability of occurrence (P) and exposure (E). The relevance of consequences and the probability of occurrence are assessed in the same way as in the assessment of threats with the use of risk evaluation. The determination of exposure starts from the predetermined fact that the time horizon (T) of analysis is 10 years (from the beginning of 2011 to the end of 2020). This datum is inserted into the relation (4) together with the outcomes of analysis aimed at the duration of threats. Then the exposure to individual threats is calculated. The outcomes are shown in Table 8.

The acquired values of all three criteria are inserted into the multicriterial matrix and the levels of risks (R) of individual threats are calculated (see relation 5) and prioritized from the most to the least relevant. The outcomes are shown in Table 9.

Table 8. The calculation of exposure to the analyzed threats

Threats	Duration of threat	Calculation of exposure	No of points
T_A	10 years (2011-2020)	$E = 10 \cdot 10^{-1} \cdot 100 = 100\%$	5
T_B	10 years (2011-2020)	$E = 10 \cdot 10^{-1} \cdot 100 = 100\%$	5
T_C	3 years (2011-2013)	$E = 3 \cdot 10^{-1} \cdot 100 = 30\%$	2
T_D	3 years (2011-2013)	$E = 3 \cdot 10^{-1} \cdot 100 = 30\%$	2

Table 9. The example of particular use of multicriterial matrix

MATRIX		assessment criteria			R	Order of Threats
		C	P	E		
Threats	T_A	2	5	5	4,00	1.
	T_B	4	2	5	3,66	2.
	T_C	3	1	2	2,00	4.
	T_D	4	1	2	2,33	3.

MULTICRITERIAL ASSESSMENT OF THREATS (OPPORTUNITIES) WITH REGARD TO THE SCALES OF CRITERIA

The previous use of multicriterial assessment assumes that all the applied criteria are equal, i.e. they have the same coefficient of relevance. However, it will be necessary to take the scales of individual criteria into account if we want to achieve a more accurate outcome. It is a numerical expression of their relevance, or, the accuracy of estimate, or predictability. The general rule[5] will be applied that the more relevant the criterion is (or, more precisely, the more relevant it is considered to be by a decision maker) the higher its scale is. Lower scales, on the contrary, will be assigned to less relevant criteria.

The scales of criteria may be determined by various methods, e.g. by pair comparison, 100 points method, and Saaty´s method. However, in case there is a small number of the assessed criteria (3 criteria in our case, i.e. the relevance or attractiveness of consequences, the probability of occurrence and exposure) it is reasonable to use the method of direct determining of criteria scales with the help of the point scale. The point scale of lower discrimination ability from 1 to 5 has been used for determining the scales, where 5 points are the most significant, while 1 point is of little significance. The outcome of assessing the relevance of criteria based on the graders´ preferences is shown in Table 10.

After determining the scales of criteria it is possible to continue with the multicriterial assessment. The procedure will be similar as in the previous case, i.e. the acquired values of all criteria will again be inserted into the Matrix and the final levels of risks (benefits) will be calculated for individual threats (opportunities). The calculation is different from the previous one in considering the determined criteria scales, which multiply individual criteria. The final value of risk (opportunity) will be calculated as a sum of weighted values of individual criteria (see relation 6). The threats (opportunities) of the highest values are considered to be the most relevant (attractive).

$$R = C \cdot S_C + P \cdot S_P + E \cdot S_E \tag{6}$$

Table 10. The determining of criteria scales with the help of point scale

Criterion	C (A)	P	E	Sum
No of points	5	5	3	13
Standard scales	0,385	0,385	0,230	1

where:

R represents the assumed level of risk of a given threat

C represents the relevance of consequences of the assessed threat

P represents probability that the assessed threat occurs

E represents exposure to the assessed threat

S represents standard scales for individual criteria.

Example: We will continue with the previous example to demonstrate the multi-criterial assessment of threats with regard to the criteria scales. At first the standard scales (see Table 10) are added to the acquired values of criteria (C, P, E) and thus the criteria will achieve different relevance. After that all data are inserted into the multicriterial matrix and the risks for individual threats are calculated. The final step is aimed at prioritizing the threats according to the final values of risks, i.e. the threats of the highest values are the most relevant (see Table 11).

After the multicriterial assessment is finished and the threats are prioritized according to the levels of risks the threats are recorded in an appropriate form, similarly as in the case of strengths and weaknesses. After that the threats are prioritized according to the achieved levels of risks (from No 1 for the threat of the highest level of risk to No X for the threat of the lowest level of risk). The threats in our case study are prioritized according to their levels of risks as follows:

1. Monitoring of network
2. Alteration of sent data
3. Overloading of information system
4. Insertion of disinformation into the information system.

Table 11. The example of multicriterial matrix used with regard to criteria scales

MATRIX		assessment criteria			R	Order of Threats
		C	P	E		
THREATS	T_A	$2 \cdot 0,385 =$ 0,77	$5 \cdot 0,385 =$ 1,925	$5 \cdot 0,23 =$ 1,15	3,85	1.
	T_B	$4 \cdot 0,385 =$ 1,54	$2 \cdot 0,385 =$ 0,77	$5 \cdot 0,23 =$ 1,15	3,46	2.
	T_C	$3 \cdot 0,385 =$ 1,155	$1 \cdot 0,385 =$ 0,385	$2 \cdot 0,23 =$ 0,46	2,00	4.
	T_D	$4 \cdot 0,385 =$ 1,54	$1 \cdot 0,385 =$ 0,385	$2 \cdot 0,23 =$ 0,46	2,39	3.

SOLUTIONS AND RECOMMENDATIONS

Every analyst must ensure that the outcomes of assessment are the most objective. This is true also in case of assessing the security of organization security systems through SWOT analysis. The level of subjectivity in the analytical process is minimized by implementing the above presented methods, which increases the quality of the acquired information. Objective outcomes help the organization to have a more precise overview of the prospects of further development.

The assessment of risks and benefits is a suitable way of acquiring the objective outcomes in the assessment of external factors. It is based on the assessment of two basic criteria, the relevance or attractiveness of consequences and the probability of their occurrence. The matrix of multicriterial assessment may also be used for the assessment of threats and benefits. The advantage of this type of assessment is in implementing more criteria, the predictability of which may be considered through standard scales. On the other hand there are also some disadvantages – it is time consuming, and criteria may be selected and assessed inappropriately. At the same time it has to be stated that theoretical knowledge and, ideally, experience are the necessary prerequisites for the implementation of these methods.

The implementation of SWOT analysis is fundamentally based on specific methods and tools, which are implemented in the identification of external and internal factors as well as in their assessment. Therefore it is suitable to analyze the possibilities of implementing the methods and tools for specific types of organizations in individual phases of SWOT analysis. The future research should be aimed at discovering the possibilities of other suitable semi quantitative and quantitative methods of assessing the strengths, weaknesses, opportunities and threats. The research aimed at discovering the purpose of implementing SWOT analysis in individual organizations of the private and public sector could also provide interesting information.

ACKNOWLEDGMENT

Language correction was done by Jiří Dvořák from University of Defence, Czech Republic. This chapter has been developed within the project of the Ministry of Interior of the Czech Republic, filed under VF20112013019code and entitled 'Objectification of Threats and Risks of Equipments for the Production and Transmission of Electricity'.

REFERENCES

Fotr, J. (2006). *Managerial decision making.* Prague, Czech Republic: Ekopress.

Grasseova, M. (2006). The implementation of SWOT analysis in long-term planning. *Defence and Strategy, 2*(6), 48-55. ISSN 1214-6463

Grasseova, M., Dubec, R., & Horak, R. (2008). *Procedural management in public and private sectors.* Brno, Czech Republic: Computer Press.

Grasseova, M., Dubec, R., & Rehak, D. (2010). *The analysis of enterprise on manager's hands: 33 the most frequently applied methods of strategic management.* Brno, Czech Republic: Computer Press.

Rehak, D., Dubec, R., & Grasseova, M. (2008). *Assessment of external environmental elements within SWOT analysis utilizing the evaluation of risks and benefits.* Paper presented at the Symposium on Risk Analysis/Management Cybernetics/Economics (19th International Conference on Systems Research Informatics & Cybernetics), Baden-Baden, Germany.

Rehak, D., & Dvorak, J. (2010). Risk catalogue as a software tool for supporting the business continuity planning. *Int. J. Business Continuity and Risk Management, 1*(2), 187-196. ISSN 1758-2164

Rehak, D., Dvorak, J., & Grasseova, M. (2009). Principles, framework and process of risk management. In J. Navrátil & J. Barta (Ed.), *International Conference Security Management and Society* (pp. 364-376). Brno, Czech Republic: University of Defence. ISBN 978-80-7231-653-3

Tomecek, P. (2008). The methods of computer attacks in the information warfare. In P. Hruza & P. Tomecek (Ed.), *The 2nd International Conference on Advanced and Systematic Research.* (pp. 79-82). Zagreb, Croatia: Faculty of Teacher Education of the University of Zagreb. ISBN 978-953-7210-14-4

KEY TERMS AND DEFINITIONS

Assessment of Risks and Benefits: The assessment of risks (benefits) is carried out by determining the relevance (attractiveness) of consequences of the analyzed threats (opportunities) in the assessed area of organization and defining the probability of threats affecting the given area.

Attractiveness of Consequences: The attractiveness of opportunity consequence is the value derived from the amount of potential benefits caused for the organization by taking advantage of such an opportunity.

Multicriterial Assessment: The multicriterial assessment is based on the assessment of factors of the analyzed area (e.g. threats and opportunities) with the help of predetermined criteria (e.g. the relevance of consequences, the probability of occurrence or exposure).

Opportunities: Factors from the external environment of organization, which may be used by an organization to increase its effectiveness and efficiency.

Probability: Probability is the value, which expresses the degree of predictability that a threat or an opportunity will affect the assessed area of the organization.

Relevance of Consequences: The relevance of consequences is the value derived from the amount of potential losses caused for the organization by threats.

Strengths: The internal factors of organization, which have positive effect on its effectiveness and efficiency.

SWOT Analysis: SWOT analysis is a method of strategic analysis of the initial state of organization and/or its part generating the alternatives of strategies on the basis of internal analysis (strengths and weaknesses) and external analysis (opportunities and threats).

Threats: Factors from the external environment of organization, which may threaten the effectiveness and efficiency of the organization.

Weaknesses: The internal factors of organization, which have negative effect on its effectiveness and efficiency.

ENDNOTES

[1]   Grasseova et al. (2010, pp. 304).
[2]   The term attractiveness of opportunity consequence shows the extent to which the given opportunity is beneficial and applicable for the analyzed group of processes.
[3]   Rehak et al. (2008).
[4]   Rehak et al. (2009), Grasseova et al. (2010) and Rehak and Dvorak (2010).
[5]   Fotr (2006) and Grasseova et al. (2010).

Chapter 8
An E-Support Firm's Response to Local E-Readiness and the Global E-Business Environment

John Effah
University of Ghana Business School, Ghana

Ben Light
University of Salford, UK

EXECUTIVE SUMMARY

The purpose of this study is to understand a small e-support firm's response to the local e-readiness and the global e-business environment in a developing context. E-Support firms provide Web development and consultancy services to user organizations, assisting them in their uptake and maintenance of their Internet applications. Within the e-readiness research area, little is known about e-support firms, particularly in connection with their interaction with their local and the global e-business environment. As yet the emphasis on e-readiness studies has been at the national level. Nevertheless, the e-support sector is very significant in the successful adoption and diffusion of the Internet and related applications in any economy. It is thus important to understand how such firms relate to their e-business environments.

DOI: 10.4018/978-1-61350-311-9.ch008

That said, this study draws on the interpretive case study of a small e-support firm in Ghana, a developing context, to investigate the firm's response to the e-readiness level of the local and the global e-business environment. Findings show that the firm could employ resources from the global environment to address most of the infrastructural challenges posed by a relatively poor local e-readiness context. However, its attempt to transfer advanced e-business technologies from the global e-business environment to the local e-business context did not succeed. This chapter offers implications for practice and research concerning the notion of reconciling local and global e-business environments in the small e-support sector.

INTRODUCTION

It is widely accepted that the Internet and the World-Wide-Web have become important technologies across our social and economic lives. They are now widely available in our workplaces as well as in our homes. For these technologies to successfully play their significant role in society and organizations, a strong e-support sector that can provide the necessary technical support for user organizations and individuals who cannot develop, implement and maintain their own systems is indispensable. Molla and Licker (2005) underscores the significance of such e-support sector for the successful adoption and diffusion of e-business/e-commerce, particularly in developing countries where ICT infrastructure and skills are relatively poor and limited. With a strong e-support sector, user organizations can concentrate on their core businesses and outsource their technical needs to specialist ICT firms.

Despite the significance of e-support firms in the both mature and emerging digital economies, research on them within the e-readiness literature is limited. As yet, the dominant focus on e-readiness research has on the national level (e.g, Bui, Sankaran, & Sebastian, 2003; Ifinedo, 2005). The e-support area with its relationship to local and global e-business environment is significantly under explored. Given its significance in the adoption and diffusion of Internet and Web-based technologies in organizations and society, it is important that the sector receives adequate research attention as an important component of e-readiness research in every country. More-over, given their supportive role, e-support firms need to understand not only their local e-business environment but also developments in the dynamic global e-business context. To do this, they need to be guided by appropriate research findings that can provide relevant information on the e-readiness of the local business environment and situations in the global e-business context. That said, this study draws on the interpretive case study approach (Klein & Myers, 1999; G. Walsham, 1995; 2006) involving a small e-support firm in Ghana, an emerging digital economy in a developing context, to understand the firm's reactions to the e-readiness state of the local e-business context and developments in the global e-business environment.

The rest of the chapter is organized as follows. In the next two sections, prior literature on general e-business support environments and the local e-business context are examined. Following this, the methodology followed to conduct the study is presented. After this, the description and analysis of the case study is provided. This is followed by the discussion of the main findings. The chapter concludes with discussion on the research contribution and implications for practice and for future research.

E-Readiness and E-Business Support Sector

The e-readiness research area has received much research attention (e.g, Bui, et al., 2003; Choucri, Maugis, Madnick, & Siegel, 2003; Ifinedo, 2005; Rodriguez-Abitia, Vidrio, & Montiel-Sanchez, 2004; The Bridges Organization, 2001). However, the dominant focus has so far been at the national level. Findings from country level studies can be useful to governments and policy-makers for comparison with e-readiness of other countries in order to identify areas in need of improvement (Jerman-Blažič, 2008). Yet, their direct value to organizations is limited. It is argued that findings from country level e-readiness research have had limited positive impact on firms and the business sector (Dada, 2006; Jerman-Blažič, 2008). Ifinedo (2005) points out that e-readiness research at the national level has failed to provide complete and relevant information to support decision-making at the organizational level.

Choucri et al. (2003) also question the practical value of e-readiness findings to individual firms. Jerman-Blažič (2008) points out that findings from national e-readiness studies are often fraught with ambiguities and inaccuracies. They criticize extant e-readiness research for using the same assessment criteria across countries, without accounting for contextual differences. They argue that such practices render the findings less useful and practical for the business communities in varied contexts. Dada (2006) also question why higher national scores have not been associated with higher organizational e-readiness and e-business adoption. For example, Purcell and Toland (2004) found that despite increasing access to the Internet in Samoa, business engagement with the technology remains limited. They attribute this inconsistency to the failure of e-readiness research to include relevant organizational factors.

The above discussions suggest that e-readiness research needs to move beyond national macro factors to include relevant business issues. Jerman-Blažič (2008) calls for frameworks that incorporate relevant organizational factors. Molla and Licker (2005) point to the need for e-readiness research to extend to the organizational level. They propose a more inclusive e-readiness framework that integrates the national, sectoral and organizational levels. It is believed that application of such approaches can make findings from e-readiness research more relevant to individual industries and organizations (Singh & Byrne, 2005). For example Barua et al. (2004) note that e-readiness for customers and suppliers are directly associated and linked to that of industries. As Molla and Licker (2005) points out, e-readiness research need to be context sensitive and pay more attention to specificities in different organizational environments.

As yet the e-support factor has not been given the necessary attention in e-readiness research. Molla and Licker (2005) identify ICT support as one of the most important sectors in determining organizational e-readiness. Without a strong e-support sector, organizational uptake of e-business technologies will remain limited. In most developing environments, the e support sector is hardly ready to support the local e-business industry. This has been reported as one of the reasons for the low organizational e-readiness in most developing countries (Richard Boateng, Hinson, Heeks, & Molla, 2008; Kapurubandara & Lawson, 2008; Molla & Licker, 2005). The above discussions support the need to take the e-support sector serious in e-readiness research.

The E-Business Environment in Ghana

Internationally, Ghana is hailed as one of the most politically and economically stable nations in Africa. In 1995, it became the first country on the continent to connect to the Internet (Foster, Goodman, Osiakwan, & Bernstein, 2004; Saffu, Walker, & Hinson, 2008). Ghana's Internet service and café industries began to grow from the mid-1990s (Foster, et al., 2004). The period also marked the beginning of e-business/e-commerce in the country. Ghana's current internet penetration stands at 4.2% (International Telecommunication Union, 2009). Although this is below the continent's average of 5.2%, it is way below the real access level as majority of users are not direct subscribers. Rather, the majority access the Internet from Cyber cafés which are found to be far cheaper than direct subscription (Foster, et al., 2004; Fuchs & Horak, 2008). Moreover, with the introduction of broadband and mobile internet services, many people have begun to subscribe to the Internet from their homes and their mobile handsets. In recent years, mobile technology penetration in Ghana has been one of the fastest globally (Frempon, Essegbey, & Tetteh, 2007).

Despite these positive developments, advanced business use of the Internet and the web has remained basic and often limited (R. Boateng, Heeks, Molla, & Hinson, 2009; Foster, et al., 2004). The use for communication and marketing through e-mail and brochure-ware far outstrip transaction based applications (R. Boateng, et al., 2009; Mbarika, Okoli, Byrd, & Datta, 2005). For example, in a study on Internet use in the hotel and tourism industry, Hinson and Boateng (2007) reported that advertisements and communication remain the dominant use among firms which hardly employ e-commerce and e-payment technology. The country's e-business environment is also fraught with a number of challenges. To date, e-commerce transactions continue to be settled through cash or cheque instead of online payment (R. Boateng, et al., 2009; Effah & Light, 2009a, 2009b; Mbarika, et al., 2005). Credit/debit card facilities for online payments are largely unavailable. Locally issued debit cards are often limited to ATMs and EPOS use. It is noted that the lack of online payment infrastructure and perceived high Internet fraud hinder the diffusion of advanced e-business technologies (Foster, et al., 2004). The West Africa Region where Ghana is located has been identified with high Internet risk. The region is also the home to Nigeria, a country globally ranked as the highest nation in Internet fraud (Adeyeye, 2008). As a consequence, the financial sector in the region has been hesitant and overly cautious to promote online payment systems.

Another hindrance to the successful development of the country's e-business environment is the limited availability of advanced e-business skills and competence (R. Boateng, et al., 2009). Foster et al.(2004) report that local Universities are not adequately resourced to train the required highly skilled ICT professionals. The few available experienced professionals often migrate to the developed world to for better working conditions. The situation has created a vacuum and a very high demand for e-business technology professionals. Those returning from the developed world with high experience and competence are too expensive for the local small firms. Such professionals often prefer the large and international organizations particularly within the financial and telecommunications industries which can assure them of higher remuneration (R. Boateng, et al., 2009). SMEs therefore have difficulties in recruiting and retaining highly skilled ICT professional with e-business expertise.

A further barrier to the local e-business environment is the high cost of Internet services (Foster, et al., 2004; Hinson & Boateng, 2007). Compared to global trends, Internet use in Ghana is expensive. It costs an average of US$1.30 an hour for browsing in a cybercafé and between US$60 and US$95 for domestic broadband subscription at the speed of 256kps (Hinson & Boateng, 2007). This is far beyond what most Ghanaians and small firms can afford. Moreover, telecommunication and Internet services are often unreliable and unreasonably slow. Frequent and unannounced electricity cuts further interrupt the services, forcing firms to fall on alternate power supply (Foster, et al., 2004). As a result, a number of firms opt to host their Websites abroad (Foster, et al., 2004) especially in the UK and USA.

Research Approach

The study forms part of a larger research project to understand e-support firms' reaction to dynamic developments in local and global e-business environment. The research follows the interpretive case study approach (Klein & Myers, 1999; G. Walsham, 1995). The fieldwork was conducted in two phases by the first author in remote collaboration with the second. The first phase occurred in November to December 2008. This was followed by the second in October 2009 to February 2010. The purpose of the second phase was to clarify issues that had emerged from the first phase and to gather more focused data.

We employed multiple data gathering techniques. We conducted formal interviews as well as informal discussions with management and employees. Two formal interviews were conducted with the CEO, one of the two management staff. On both occasions the other manager was not available for interview. Each interview was based on a semi-structured interview guide. The guide was flexibly designed to accommodate participants' emergent concepts and themes. Each interview lasted between 1 and 1.5 hours, was tape recorded following participant's consent, and later transcribed. We mailed the transcripts to the participant for verification. Follow-up interviews were conducted via e-mail and telephone. We collected further data from company documents and the Internet including the websites of the firms and its clients. Additional data came from the analysis of newspaper archives.

As the study is interpretive in nature, data gathering and initial analyses were conducted simultaneously. We subsequently undertook detailed analysis post data collection, as we carefully read and reflected on the field notes, frequently discussing the emerging themes and concepts between us following a similar approach in Walsham and Saha (1999). We present the key findings from the analysis below.

The Case of SupportCo

SupportCo (a pseudonym) is a family owned business providing e-support services to firms in Ghana and other parts of Africa. It is wholly owned by the CEO and his wife, who are both IT professionals and Dutch expatriates residing in Ghana. The firm was originally established in 2001 as a subsidiary of NedaNet (a pseudonym), a Web development company in the Netherlands. The CEO was an employee of NedaNet. In 2002, he was transferred to Ghana head SupportCo and relocated with his family. Later in the same year, NedaNet was acquired by GloNet (a pseudonym), another Web development company in the Netherlands. However, GloNet had no interest in SupportCo. Therefore the CEO and his wife acquired it and began to manage it themselves, the CEO taking charge of software development and administration with his wife responsible for client project management.

The couple had extensive experience in ICT, having worked in varied countries across Europe, the United States, and Africa. By 2010, the number of employees stood at 14 excluding the two managers. These include technical staff consisting of Web developers and editors, programmers, system analysts, designers and administrative staff including one each for marketing, sales, and accounting. SupportCo's original focus was solely on Web design, development and maintenance for local clients. Since the management found the dominant project-based income source to be irregular, they decided, from 2006 to diversify revenue sources to include additional service lines such as Website hosting, application service provision, search engine optimization, and intelligent e-mail systems. E-Support's key strategy has been to monitor the global e-business environment to identify innovative technologies and transfer them to the local environment. The rest of this section discusses e-readiness of the local environment and the firm's response the significant factors that shape its operation and strategies.

The IT Labour Market

According to the CEO it is very difficult to attract and retain employees with the right expertise in Web design, analysis and programming skills in Ghana. Comparing his experience in the country with that in Europe, the United States and other parts of Africa, he lamented that that of Ghana is perhaps the worst so far. Moreover, SupportCo's labour turnover has been high over the years. In 2005 alone, the firm lost as many as 13 employees and had to hire 16 more to meet its technical employee requirements. To address the issue, the firm had to resort to outsourcing part of its job to Web development experts in Europe and the United States. The CEO claimed that: "it is cheaper for us to outsource [jobs] to developers in New York than to do them in Ghana". He however added that not all services can be outsourced, particularly those requiring face-to-face interaction with clients.

Local Infrastructure

The supporting infrastructure for the Internet and related technologies in Ghana is relatively poor and more expensive. The CEO lamented on the quality and high cost of infrastructure in the country:

"I mean the infrastructure in Ghana is still a very big problem. In Europe you will get 2mgb/s connection in your home for $50 or less. If you want to get the same connection here you will pay $150 or more. Aside the cost, telecommunication infrastructure is both slow and unreliable. Until recently, it was difficult to get access to broadband services".

Due to the weak infrastructure, users spend much more time accessing Websites from Ghana than in Europe or the United States. The high costs also hinder the adoption and widespread diffusion of the technology in organizations, especially with small firms which form the majority of organizations in the country.

Until 2006, SupportCo hosted its website and those for its clients in Europe and the United States. Although this is far cheaper than getting them hosted locally, the offshore hosting strategy creates the problem of latency, the increase in accessing time even for local Websites. Moreover, given the weakness of the local currency to the Dollar due to the unfavourable exchange rate, organizations inexperience increases in hosting following the falling of the local currency to the Dollar as firms charge in the local currency but pay for hosting in dollars. To address this and also as part of the strategy to ensure a regular source of income, the management decided, from 2006, to offer local hosting services. The CEO believes that things will improve after the implementation of a local Internet Exchange in the country. The firm is therefore continuing to assess the level of improvement in the local Internet infrastructure. When it is in place, it will host all local websites in Ghana.

Unreliable access to electricity remains another challenge for the smooth operation of SupportCo and its clients. They experience frequent power cuts particularly during the dry season. In Ghana, electricity is mainly generated from hydro-power. Hence the country suffers from inadequate supply during low periods of rainfall. During such periods, SupportCo incurs additional costs in diesel to power a stand-by generator.

Online Payment Facilities

To date, there are no online payment services as such to support e-commerce in Ghana. The banks do not issue credit/debit cards to be used online. Online credit cards such as Visa and MasterCard are not in common use. Consequently, there is a lack of a coherent Internet payment system. Intelligent cards issued by the banks are also limited to use at ATMs and EPOSs. Although this remains a big hindrance for SupportCo in dealing with customers, suppliers, and business partners, it has had no other option than to operate in cash or cheque. Given their experience in the West where online payment is common, the management of SupportCo are often frustrated in conducting all transactions in cash or by cheque. For now, they can only hope that the situation will change. They have not been able to identify any better alternative for online settlement.

Local Demand for Web Technologies

Most firms in Ghana are yet to embrace e-commerce as an alternative business channel. Thus far, most business continues to employ the traditional channel. Those employing the Internet and the Web tend to limit their use to basic applications such as e-mail and advertising. Moreover, rather than use professionals such as SupportCo, they would prefer to use friends and relatives to develop basic websites which they find cheaper. Therefore, despite SupportCo's expertise and competence in Web development, the demand for services remains low. The situation has negatively affected the firm's cash flow and profitability. The CEO described the firm's financial performance as follows: "Quite frankly we are not making profit. Every year we are just breaking even or making a little bit of a loss or a little bit of a profit." Management attribute the financial difficulty to project-based income such as from website development which is found to be irregular in Ghana. To address this, the firm has initiated strategies to diversify into areas that can ensure a regular source of revenue.

SupportCo's Strategic Response to the Local E-Readiness Environment

This section presents SupportCo's response to developments in both the local and the global e-business environments. Due to the low demand for SupportCo's core services for Website development and the irregular source of income, the management decided, in 2007, to diversify into new areas that could ensure a regular flow of income. The CEO commented that:

"90% of our income was from web-development but the disadvantage of web development is that it is a project based-income [an irregular source]. So we very quickly realised that we needed to increase our recurring revenue".

He was optimistic that the situation could improve given the new strategic move and the perceived increase in the number of Internet users in the country. He had an increasing trend in the hits per day on the website of one of SupportCo's client, GhanaWeb, the most popular media portal in Ghana. GhanaWeb provides up-to-date news and a wide range of information on Ghana. It also offers discussion forums for the people of Ghana and foreigners on topical issues of the country. According to the CEO, this website receives well over 122,000 unique visits per day. To him, this points to a possible increase in the number of Internet users in Ghana which hopefully could be translated into demand for Internet and related services by individuals and local firms.

In order to diversify SupportCo's income sources to counter the negative and irregular cash flow problem, the management decided to import cutting edge technologies from the global environment to the local business community. First, they introduced online application services where clients could pay to use applications developed and hosted by SupportCo rather than buy it themselves. This was followed by the introduction of a search engine optimization service where clients' websites competing for online visitors' attention could emerge on top of other websites during an Internet search by potential customers. Finally, the transferred intelligent e-mail services to the local business community. This service, according to the CEO, could be used by firms to analyze and monitor incoming and outgoing e-mails for management decision making and control purposes. Unfortunately, the local business community's response to these cutting-edge innovations from the global business environment has been below the expectation of the management of SupportCo. So far, they have come to believe that local organizations do not appear to be ready for such innovations despite their potential usefulness. The management became confused of what to do next since the strategy appeared not to be the solution for the negative cash inflow and irregular source of income. They wonder whether they should shut down and return to Europe or continue to hope for a better local e-readiness environment.

DISCUSSION

The purpose of the study has been to investigate the response of an e-support firm in a developing context to the local e-readiness business environment. Findings from the study show that SupportCo managed to address some of the challenges that emerged from its interaction with the poor resource status of the local e-business environment. First, to address the non-availability and high cost of local expertise, it outsourced some of its jobs to the developed world where it found relatively higher ICT skills availability at lower cost. With this, the firm could meet the problem of limited ICT labour and skills in the local e-business environment. Next, to address the problem with the relatively poor but expensive local hosting services, it managed to host websites for its client's off-shore, particularly in Europe and the United States.

This was also found to be far cheaper. However, two problems emerged from this approach. First, the fact that the local currency was weaker to the US Dollar which was used to quote for off-shore hosting fees, meant that local firms could experience exchange losses. Whilst local firms earn their revenues in the local currency, they had to pay for hosting services in a stronger currency. Given the frequent unfavourable exchange rates against the local currency, local firms could incur losses resulting from currency fluctuations. Another problem, which was noted by the CEO of SupportCo, is the increase in the latency time to access local websites from other countries. Clearly, if all local websites were hosted locally, it should be faster to access such websites given the proximity of local web servers. However, with the off-shore hosting, local Internet users would need to access servers residing in other countries before gaining access to local websites. This could lead to delays which could have been avoided with local hosting.

Another infrastructural difficulty is the unreliable access to electricity, particularly during the dry season as the country's power supply is subject to the weather. This means organizations have to incur additional costs to acquire stand-by generators and power them with diesel. This alternative supply has been more expensive. Despite the firm's effort to address the above challenges, there were some other difficulties to which it could not find any solution. The most obvious is the lack of an online payment system which could enable it to settle and receive payments electronically. For now it has to cope with the situation by conducting all transactions with its suppliers and customers in cash or by cheque. Hopefully, given the gradual diffusion of online banking systems in the country, the situation may improve in the future so that payments can be done online.

Although it could be said that SupportCo has done well to address the relatively poor e-readiness status of the local business environment, its response to addressing the negative cash flow and irregular source of income appears to have failed. Clearly, the firm has been reactive rather than proactive to the real business needs of the local business community. The adopted strategy to diversify its income sources by importing cutting edge e-business innovations failed to meet the needs of the local business firms. The local business community appears not to be e-ready for such advanced technologies. Perhaps SupportCo's strategy could have worked if it had identified local needs and looked for innovative technologies from the global business environment to meet such needs. Wholesale transfer advanced technologies to the local business environment would not necessarily be adopted by local businesses if they do not meet their current needs.

As prior studies on Ghana's e-business environment show (R. Boateng, et al., 2009; Foster, et al., 2004; Hinson & Boateng, 2007; Mbarika, et al., 2005) the local business community is not yet ready for advanced e-business technologies such as online application services, intelligent e-mail system and search engine optimization that SupportCo decided to offer. So far, the use of e-business technologies in the country is largely limited to basic applications such as e-mail and brochure-ware for online advertisements (Hinson & Boateng, 2007). Although management is optimistic that user hits on GhanaWeb, the most popular Web site, are increasing and therefore Internet penetration in the country will grow, Foster et al.(2004) argues that an increase in website visits access in Ghana does not necessary translate into high business use. This is supported by the findings in Purcell and Toland (2004) that, in a typical developing country context, increasing access to the Internet does not necessarily generate high business use. Moreover, Ghana is noted to be highly socially networked and people communicate a lot with friends and relatives. The higher diffusion of mobile phones far ahead of computers attests to this culture.

The findings from the study challenge the common rhetoric that the Internet is creating a global village. As Molla and Licker (2005) points out, marked differences exist between e-business environments in the developed and the developing world. Therefore merely scanning the global environment for advanced innovative technologies and transporting them to the local environment such as SupportCo may not be feasible. The firm ought to have assessed the readiness of the local environment for technologies to meet such needs.

SUMMARY AND FINDINGS

This study investigated the reaction of a small e-support firm in Ghana to the relatively poor local e-readiness environment and developments in the global e-business context. The firm succeeded with recourse to the global e-business environment to address the challenges it faced in a limited local e-readiness context. It was able use outsourcing and off-shore hosting in the developed world to overcome the limited skills and infrastructure resources in Ghana. However, failure to understand e-readiness of the local business community to the specific applications it transferred from the global environment to the local environment turned its strategy to reverse the poor financial performance into a failure. Clearly the local firms were not ready for the advanced technologies it attempted to transfer to them. The firm ought to have analyzed the immediate needs of the local organizations in order to identify appropriate technologies from the global environment to meet such needs.

These findings suggest some implications for practice and for research. For practice, e-support and technology vendors need to pay attention to the business needs of local firms and transfer technology that would meet such needs. Failure to pay attention to local needs may lead to poor acceptance of technology transferred from the global e-business market, regardless of the level of sophistication and its usefulness in the global environment.

For research, e-support firms need to respond to two different e-business environments - local and global. Developments in the two environments differ owing to differences in complexity and dynamic factors. Given their limited resources, e-support firms need to be guided on how to assess the two environments to reconcile local technology needs and developments in the global e-business environment. If e-readiness research can be useful to local e-support firms, it would be important to incorporate factors that can provide information on the e-readiness of user organizations. Future research can focus on developing useful models and frameworks to guide e-support firms to assess local technology needs and developments in global e-business environments.

REFERENCES

Adeyeye, M. (2008). E-commerce, business methods and evaluation of payment methods in Nigeria. *Electronic Journal of Information System Evaluation, 11*(1), 1–6.

Boateng, R., Heeks, R., Molla, A., & Hinson, R. (2009). E-commerce in Ghana: Where we are and where we are headed. In Hinson, R., Boateng, R., & Mbarika, V. A. (Eds.), *E-commerce and customer management in Ghana* (pp. 23–59).

Boateng, R., Hinson, R. E., Heeks, R., & Molla, A. (2008). E-commerce in least developing countries: Summary evidence and implications. *Journal of African Business, 9*(2). doi:10.1080/15228910802479919

Bui, T. X., Sankaran, S., & Sebastian, I. M. (2003). A framework for measuring national ereadiness. *International Journal of Electronic Business, 1*(1), 3–22. doi:10.1504/IJEB.2003.002162

Choucri, N., Maugis, V., Madnick, S., & Siegel, M. (2003). *Global e-readiness - for what?* MIT Sloan School of Management.

Dada, D. (2006). E-readiness for developing countries: moving the focus from the environment to the users. *Electronic Journal of Information in Developing Countries, 27*(6), 1–14.

Effah, J., & Light, B. (2009a). *Beyond the traditional 'SME challenges' discourse: A historical field study of a dot.com failure in Ghana.* Paper presented at the UK Academy for Information System (UKAIS) 14th Annual Conference.

Effah, J., & Light, B. (2009b). *Understanding SME E-business challenges in developing countries: The case of dot.coms in Ghana.* Paper presented at the British Academy of Management (BAM) Conference.

Foster, W., Goodman, S., Osiakwan, E., & Bernstein, A. (2004). Global diffusion of the Internet IV: The Internet in Ghana. *Communications of the Association for Information Systems, 13*(1), 654–681.

Frempon, G., Essegbey, G., & Tetteh, E. O. (2007). *Survey on the use of mobile phones for micro and smal business development: the case of Ghana.* Accra, Ghana: CSIR-Science and Technology Policy Research.

Fuchs, C., & Horak, E. (2008). Africa and the digital divide. *Telematics and Informatics, 25*, 99–116. doi:10.1016/j.tele.2006.06.004

Hinson, R., & Boateng, R. (2007). Perceived benefits and management commitment to e-business usage in selected Ghanaian tourism firms. *Electronic Journal of Information Systems in Developing Countries, 31*(50), 1–18.

Ifinedo, P. (2005). Measuring Africa's e-readiness in the global networked economy: A nine-country data analysis. *International Journal of Education and Development using Information and Communication Technology, 1*(1), 53-71.

International Telecommunication Union. (2009). *Internet usage statistics for Africa.* Retrieved January 24, 2009, from http:// www.internetworldstats.com/ stats1.htm#africa

Jerman-Blažič, B. (2008). Web-hosting market development status and its value as an indicator of a country's e-readiness. *Telecommunications Policy, 32*, 422–435. doi:10.1016/j.telpol.2008.04.007

Kapurubandara, M., & Lawson, R. (2008). Availability of e-commerce support for SMEs in developing countries. *The International Journal on Advances in ICT for Emerging Regions, 1*(01), 3–11.

Klein, H. K., & Myers, M. D. (1999). A set of principles for conducting and evaluating interpretive field studies in Information Systems. *Management Information Systems Quarterly, 23*(1), 67–93. doi:10.2307/249410

Mbarika, V. W. A., Okoli, C., Byrd, T. A., & Datta, P. (2005). The neglected continent of IS research: A research agenda for Sub-Saharan Africa. *Journal of the Association for Information Systems, 6*(5), 130–170.

Molla, A., & Licker, P. S. (2005). Perceived e-readiness factors in e-commerce adoption: An empirical investigation in a developing country. *International Journal of Electronic Commerce, 10*(1), 83–110.

Purcell, F., & Toland, J. (2004). Electronic commerce for the South Pacific: A review of e-readiness. *Electronic Commerce Research, 4*, 241–262. doi:10.1023/B:ELEC.0000027982.96505.c6

Rodriguez-Abitia, G., Vidrio, S., & Montiel-Sanchez, C. (2004). *Assessing the state of e-readiness for small and medium companies in Mexico: A proposed taxonomy and adoption model.* Paper presented at the AMCIS 2004 Proceedings. Paper 78.

Saffu, K., Walker, J., & Hinson, R. (2008). Strategic value and electronic commerce adoption among small and medium-sized enterprises in a transitional economy. *Journal of Business and Industrial Marketing, 23*(6), 395–404. doi:10.1108/08858620810894445

Singh, M., & Byrne, J. (2005). Performance evaluation of e-business in Australia. *Electronic Journal of Information Systems Evaluation, 8*(1), 71–80.

The Bridges Organization. (2001). *Comparison of e-readiness assessment models.*

Walsham, G. (1995). Interpretive case studies in IS research: nature and method. *European Journal of Information Systems, 4*, 74–81. doi:10.1057/ejis.1995.9

Walsham, G. (2006). Doing interpretive research. *European Journal of Information Systems, 15*(3), 320–330. doi:10.1057/palgrave.ejis.3000589

Walsham, G., & Sahay, S. (1999). GIS for district-level administration in India: Problems and opportunities. *Management Information Systems Quarterly, 23*(1), 39–66. doi:10.2307/249409

Chapter 9

The Use of Collaborative Technologies within SMEs in Construction:
Case Study Approach

Vian Ahmed
University of Salford, UK

Aisha Abuelmaatti
University of Salford, UK

EXECUTIVE SUMMARY

Collaborative environments have been evolving and effectively employed in large organisations and are believed to have high potential for Small and Medium Enterprises (SMEs). This chapter shares the findings of a case study that was conducted on twelve companies in order to assess the use of collaborative environments and their adaptation approaches through interviews with senior level managers and end-users. The need for such case studies has risen from an intensive literature review which revealed that SMEs are key players within the construction industry; however, there seems to be little evidence of their utilisation of IT for collaborative learning environments. Therefore, this calls for the necessity to developing an approach blending the right combination of factors which are believed to contribute towards the improvement and implementation of collaborative environments and may affect their success.

DOI: 10.4018/978-1-61350-311-9.ch009

INTRODUCTION AND BACKGROUND

Due to its multi-organisational and geographically dispersed project nature, there are traditional collaboration requirements in construction. However, the role of IT has been overlooked in construction initiatives. In 2002 the report entitled 'Accelerating Change' was the first industry report to mention IT explicitly; in 2006 the report entitled '2012 Construction Commitments' says: "IT-based collaborative tools and communication technologies will be exploited". Yet, there is only one mention of IT in the 'Draft Strategy for Sustainable Construction' (Wilkinson, 2005). Currently, there is a gradual shift towards collaborative working and enthusiasm about the adoption of collaboration technologies that can be said to be two faces of one coin. However, still it is not satisfactory given the fact that the use of collaboration technology remains low among 99% of companies in the UK construction industry usually referred to as Small and Medium Enterprises (SMEs < 250 employees) (Barbour 2002, p.31; Barbour 2003, p.14; DTI benchmarking study 2004, p.52; Wilkinson 2005; ebusiness W@tch 2006). It is widely recognised that SMEs perish quicker than large organisations; in fact, the fairly recent Small Business Service (2004) statistics reveal that SMEs sustainability is an issue; which can be attributed partially to lack of profitability, and that profitability is linked to performance. The implementation of collaborative environments is one possible solution to improve performance among SMEs.

Previous research relates to the integration of IT in business environments in general (Underwood and Alshawi 2000; Pena-Mora et al 2002; Roshani et al., 2005; Alshawi, 2007) but the growing popularity of collaborative environments in the construction industry has, unfortunately, not been matched by parallel empirical research for SMEs. Given that SMEs deliver 52% of the construction industry's workload (DETR, 2000), it ensues naturally that they are key players in supporting large construction companies. Therefore, SMEs' good performance and survival in the industry is vital. This is the reason why the research reported in this chapter attempts to investigate ways of getting the SMEs to engage more effectively in collaboration initiatives to meet the demands of an over growing industry while increasing their overall competitiveness.

This chapter therefore looks into the key areas to focus on during collaborative environments implementation that can improve SMEs performance in collaborative working. In a study to find out the current collaborative environment implementation and collaborative working approaches covering a number of SMEs, twelve interviews were conducted in a semi-structured format with senior level managers and end-users. As a whole, this chapter concentrates mainly on presenting the results obtained from SMEs in the construction industry in the United Kingdom to explore the efficacy of different technologies for collaboration and gather information on the experiences of industry professionals during the implementation of a new collaborative environment. It is expected that more SMEs will want to learn and understand more about the technologies available in and appropriate to their home market. A discussion of this kind is useful in ascertaining using the technology collaboratively.

INFLUENCING FACTORS RELATING TO IMPLEMENTING COLLABORATION TECHNOLOGY

The literature review on collaborative environment implementations has revealed a number of issues that need to be considered with respect to failure of IT related implementations and collaborative working. These were mainly related to how it is introduced to large organisations. In an attempt to improve SMEs performance in collaboration initiatives, fifteen of the issues that are said to likely influence the success of implementing collaborative environments have been identified for further discussion. Five of these focus on organisational dimension and are interlinkedm, namely: process vision development, strategic planning, team working, decision making and perception in relation to change, risk management. Seven are socio-cultural in nature, namely: relationship, communication, empowerment, commitment, trust, mutuality and work attitudes. The other three are related to the legal aspect. These fifteen factors are classified into three alternative though complementary categories, namely: organisational, socio-cultural, and legal.

Category I: Organisational

Process vision development: process improvement and IT implementation is about redefining the company's vision. Process vision describes the future state of process and therefore links business strategies with procedures and actions (Harris and Harris 1996; Cornick and Mather 1999; Ruikar et. al, 2005; Alshawi, 2007). Therefore, SMEs have to put forward an imaginative understanding of the future processes. Indeed, introducing IT into an SME requires proper planning and careful project management effort to enable all the other key areas.

Strategic planning: Business strategy manifested in what is termed strategic planning should be underpinned by the technology strategy. In fact, the business strategy, organisation strategy and IT strategy are interrelated (Robson, 1997; Feeny and Willcocks, 1998; Alshawi, 2007). Therefore, it can be derived that when there is a change in IT structure it must be according to the IT strategy which is aligned with business strategy and organisational strategy. Where there is an IT change, it must be reflected in the organisational structure. Therefore, SMEs should keep in mind that a change in one strategy will affect the other two. Depending on the level match between the characteristics of the collaborative environment and the characteristics of the SMEs, the companies may have to carry out some changes at the project level or they may have to define a new strategic vision and carry out changes at the organisational level aligned to this vision.

Teamworking: teamwork which is organised around business and project processes is the one and only structure in the construction process that guarantees response to project delivery (Cornick and Mather, 1999; Anumba et al, 2002; Baiden et al, 2003; Bromley et al, 2003). Being the one and only structure does not imply that teamwork is entirely without its own issues. This clearly indicates that SMEs ability to overcome the challenges of full teamwworking and process standardisation is of utmost importance. The six dimensions of teamworking namely: single team focus and objectives; seamless operation without boundaries; unrestricted cross-sharing of information; creation of single and co-located team; equitable relationships, opportunities and respect for all; no blame culture (Bromley et al., 2003; Alshawi, 2007; Erdogan, 2010) assist in seeing how SMEs so far have managed to overcome the challenges of various teamworking passes. It is important to investigate and assess the extent of teamworking achieved within the construction project team in a more practical way of specific practices to indicate the achievement of full, partial, or no teamworking.

Decision making and perception in relation to change: technology brings about change at an organisatioanl and individual level, and is therefore likely to be met with resistance (Rezgui et al., 2005; McAdam and Galloway, 2005; Goudling, 2007). The resistace to technology in SMEs is a combination of resistance to the principle of collaborative working and the resistance to the adoption of the technology itself.

Risk management: underlying the collaborative environment there is a perception of risk arising from the technology adoption (Hammer and Stanton, 1995; Alshawi, 2007; Murdoch and Hughes, 2008). As such, introducing an IT enabled system into an SME requires careful management effort.

Category II: Socio-Cultural

Relationships: SMEs should make rational decisions about the types of inter-organisational relationships that they would enter into which would most benefit their company.

Communication: A common means of communication should be decided by all key participants collaborating (Rye, 1996; Holt et. al, 2000; Kitchen and Daly, 2002; Proctor and Doukakis, 2003). Specifications for the people to work on the project, the workflow of the project, and the details for the technology solution, collaboration standards and procedures should be agreed before the collaborative environment is set up.

Empowerment: Promotes self-management and collaborative team-working principles (Mumford, 1995; Arendt et al. 1995; Holt et al, 2000; Alshawi, 2007). When SMEs' empower their employees, they become more involved in deciding how work should be approached and which technologies to use. As such they will believe in and control what happens to their work processes, and will be capable of controlling those processes efficiently.

Commitment: Collaboration leaders need to ensure that all key participants are consulted as to the practices to be employed during the collaboration (Baldwin et al 1999; Thorpe et al, 2001; Weippert and ajewski, 2002). Therefore, SMEs need to be fully committed to using the new collaborative environment with buy-in and collaboration at the highest level within the participating companies.

Trust: Time and resources are needed to enable SMEs to build trusting relationships.

Mutuality: There are alternative conceptualisations of mutuality. SMEs should consider it to reflect interdependent or reciprocal patterns of subjective events.

Work attitudes: Cultural problems are brought about by employees' deep-rooted values and beliefs (Wilkinson, 2005; Alshawi, 2007; Murdoch and Hughes, 2008). As such, SMEs should reach common agreement concerning social rules to prevent problems.

Category III: Legal

Contracts: Need amendment to be appropriate for projects where IT enabled collaboration is employed (Carter et al., 2002; Wilkinson, 2005; Hassan et al. 2008). It is therefore vital that appropriate legal arrangements are put in place to support the key relationships; namely: between a software vendor/ Active Service Provider (ASP) and a client involving a Master License Agreement; between a software vendor/ASP and end-user involving an End User License Agreement; Service Level Agreement to underpin security and reliability requirements. It is also useful to specify relationships between individual members stipulating use of the technology to communicate with each other covered in project protocol documents. Not to mention managing service interruption and ASP termination.

Admissibility: Where a statement contained in a document is admissible as evidence in civil proceedings ought to be authenticated in such a manner as the court may approve (Wilkinson, 2005). At this point in time, the importance of what is termed certificate authority, digital signature, and digital notary is of utmost importance.

Ownership: Sharing project information in an electronic collaborative environment can make team members anxious about the ownership, use, and possible abuse, of the information they contribute. Such anxieties can also extend to intellectual property.

The fact of the matter is that the stand-alone implementation of collaboration technologies will not be able to enhance collaborative working unless supported by the relevant factors. Based on such assumptions, the rest of this chapter will look into exploring the challenges faced by SMEs in implementing collaboration initiatives in the construction industry, the techniques and technologies used and their effectiveness for collaboration in SMEs in the construction industry, the impact of collaborative environment initiatives on the competitiveness in SMEs in the construction industry and the means that facilitate the development of construction collaboration technologies as a core capability within SMEs.

SMEs CASE DESCRIPTION

In order to answer the research questions in the previous section, the findings from semi-structured interviews in twelve case studies of which eight SMEs selected on the following basis were included: each SME should already be engaged in collaborative arrangements; should be technology friendly, i.e. already using IT to support part of their business functions; should have a strong business focus and area of activity. Active project members involved in the implementation of the collaborative environments, whom engaged in the collaboration, experienced the difficulties and barriers, and made decisions to overcome them were considered. Large organisations were also chosen since most of the construction SMEs had referred to them. Therefore, in order to obtain the perception of the large organisations, four large companies were included in the case studies.

Profile of the Case Study Companies

A total of 16 interviewees from 12 organisations participated in the study. The number of employees in the participating companies varied from 10 - 230. The study sample consisted of 4 small and 4 medium enterprises, as well as 4 large organisations. Attempts were made to have a sample across the construction sector incorporating an assortment of professions to get a blended view of the use of IT for different purposes. Interviews with personnel across organisational hierarchy allowed some insights at organisational, project and individual levels to be gained. The case study companies are summarised in the following Table 1.

RESULTS AND MAIN FINDINGS

The aim of this section is to bring together information from the interviews that lasted approximately one hour with managers and end users to provide a summary of information found in the questions asked. The use of collaborative technologies

Table 1. Case study companies

Case No	Company type	Employees	Intervieweed personnel
Company 1	Engineering and environmental consultancy, providing support services. Core expertise is in project management, construction contracts, cost management, engineering, architecture and risk management.	SMALL	Managing Director
Company 2	A dynamic practice set up to deliver Architecture.	SMALL	Architect
Company 3	Provides clients with bespoke management services in a more personal way.	SMALL	Managing Director
Company 4	Offer a diverse range of systems capable of solving complex foundation problems, i.e. smart design solutions at all stages in the life of a project	SMALL	Design Engineer - Manager
Company 5	Urbanists in the widest sense, working collaboratively to fuse urban design and landscape architecture with ecology, environmental design, masterplanning, architecture, branding, lighting, arts, media and engineering.	MEDIUM	Director
Company 6	Offer a range of development management, project management and construction services and have the systems, processes to provide solutions to suit the clients specific requirements.	MEDIUM	Construction Manager - Document Controller
Company 7	Offer comprehensive consulting services in engineering design. Principle disciplines include civil, structural, geo-technical, and geo-environmental engineering together with complementary services such as development planning, traffic and highways engineering, conservation, and project management.	MEDIUM	Director
Company 8	Offering a network of support to projects with a group of specialist architects and technologists.	MEDIUM	Director
Company 9	The Group is organised into three business streams sharing the same values of partnering, care for environment, quality and team work. Within these business streams are the specialist expertise and know-how to deliver for clients' very specific needs.	LARGE	Senior design and build coordinator
Company 10	Specialises in delivering ambitious and innovative construction projects that deliver to both clients and the communities who use them every day.	LARGE	Project Leader
Company 11	Provider of infrastructure services specialising in the design and delivery of complex engineering projects, and provides specialist and innovative tunnelling, utilities, civil engineering, mechanical, and electrical services to the water, transport and energy sectors.	LARGE	Site Agent
Company 12	Interdisciplinary practice of architects, designers, engineers and urbanists. It combines expertise across disciplines, locations, sectors and all major building types.	LARGE	Architect Associate

to enhance collaborative working was probed. Pertinant questions were asked to explore what percentage of collaborative environments implemented in the SMEs has failed to provide the full benefits expected and investigated how the success of the collaborative environments is evaluated and whether there are any defined success criteria for this. Questions on how the collaborative environments were implemented and the factors affecting their success tried to identify whether the implementations were accompanied by any management efforts. The most and the least successful collaborative environment implementations in the SMEs within the last five years were investigated to understand the factors affecting the success. Apart from these factors, the collaborative environment implementation approaches of the company are investigated to find further success factors and failure reasons specific to the company.

The Collaborative Environments Implemented

To assess the adaptation level of collaborative environments, the interviewees were asked to categorise their implementations; namely: whether they developed a customised environment in-house by hiring either a consulting company or programmers to create a system; developed an environment by purchasing commercial web-enabled packaged software and installing it on a company's internal server; rented/leased a completely developed environment from an ASP for a usage fee, which is normally charged per project, per the amount of computer storage space required, and/or per user. Among these three options, the third was found the most common environment implemented by the interviewed companies. To determine which of the collaborative environments available to the construction industry was being used in projects, Figure 1 summarises these results. It can be seen that 4Projects is the most commonly used with Asite and BIW the next popular.

The study revealed that collaborative environments are not generally implemented in SMEs; however their use is driven by the use of larger companies. The

Figure 1. Results of the use of different collaborative environments

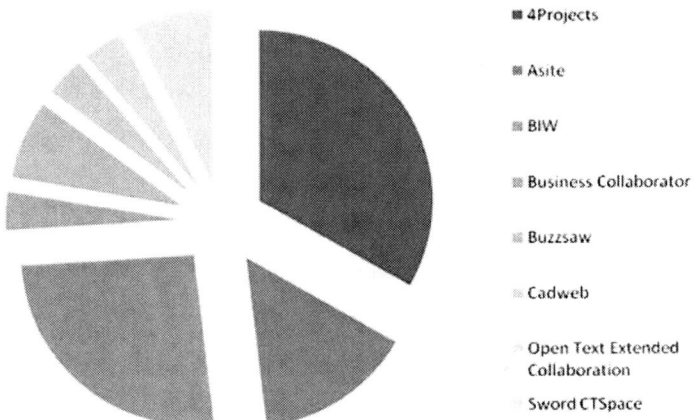

interviewees voiced concerns about unfriendly user interfaces, low speed of transfer and data security problems as the main failure reasons from the technical point of view. Respondents felt that the intuitiveness of these means that much training is needed before project participants can use them. The actual layout of them also proves difficult with some respondents commenting that there seems to be a lack of thinking by designers. Respondents would like to see simpler interfaces designed for collaborative environments. While SMEs may not be able to change the user interface of a given environment, this aspect could be enhanced with the provision of training and availability of resources to employees, they can have direct impact on the timing of end-user involvement and the quality of the training provided. As such, SMEs must be given ample training and resources to ensure optimum productivity with the implemented environment and they must be able to deal with the environment as a tool that can help their performance.

Success Level and Success Criteria of Implementing Collaborative Environments

One of the key questions asked was what percentage of collaborative environments implementation initiative has failed? This was followed by asking what percentage of collaboration technology implementation projects would not have been found financially viable if the actual success levels could have been foreseen? The results show up to 50 to 70% failure in providing the full benefits expected from the collaboration technologies implemented. Another question asked was designed to determine what are the success criteria for IT implementation, or specifically for collaborative environments implementation? With regard to the success criteria for implementations and how the companies measure the extent to which implementations satisfy these criteria, the case study revealed that they were able to achieve this success analysis perceptually through the means of tangible and intangible comparisons with cases where paper-based systems were used. The interviewees said that they were using the following questions as objective measures to find out where the cost savings mainly arise from compared to the conventional method, where they used a paper based system, namely:

- Direct tangible savings arising from on print distribution, storage and management, and the capability to track changes electronically; manifested in "How many drawings were produced?", and "How much would it have cost to produce those drawings and distribute them in paper-based form?".

- Intangible benefits arising from reduction in mistakes and re-works and by avoiding unnecessary project delays manifested in "Did we have more or less rework?", "Did we have more or less drawing revisions?", "Did we have more or less Request For Information", and "How long did it take to resolve our Request For Information?"

These indicators are not only related to the success of the collaborative environment; they are affected by many other organisational and project level factors. Therefore this perceptual analysis does not measure the success of the collaborative environment solely. These comparisons could fail in measuring the efficiency of the collaborative environment since sometimes the information is exchanged electronically without the use of the collaborative environment. Furthermore, the values calculated this way would only reflect the benefits obtained due to the automation of the communication and not necessarily the collaborative environment. Measuring the intangible benefits was found difficult by the SMEs, when there was a need to measure the benefits they either choose to do a perceptual analysis or measure the construction project instead of the collaboration technology against a number of benchmarks or key performance indicators defined at the very beginning of the project. This means that none of the perceptual analyses carried out by any of the companies managed to judge the performance accurately. They were mostly subjective and did not include the views of the end users. These analyses failed to provide results that could be used as feedback for future implementations.

FACTORS AFFECTING THE SUCCESS OF THE COLLABORATIVE ENVIRONMENTS

The interviewees were asked to name the common factors of their very successful collaboration implementations. Failure reasons were identified by each case and most of these were found to be interrelated when investigated together. Figure 2 summarises the results in order of importance. Figure 2 tells us that all the aspects listed are important to the development of a methodology that enables the effective implementation of collaborative working in the construction industry. However, there are a number of factors that really make a significant contribution to the success of implementing effective collaboration, namely: process and strategy, user involvement in decision making, risk management, commitment, empowerment, building trust in relationship, communication, teamworking, and contractual obligations. These significant factors are discussed in more detail with respect to the key dimensions for collaborative environment implementation highlighted earlier in Section Two.

Figure 2. The most important success factors

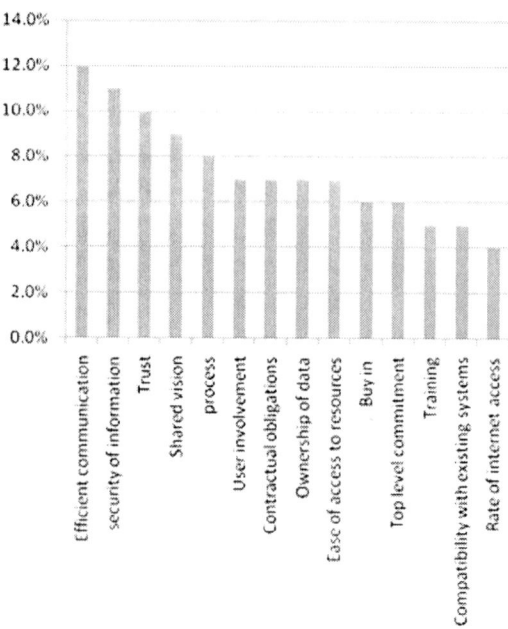

Process Vision Development and Strategic Implementation

In light of the previously published research, the collaborative environment implementation process is divided into nine steps; namely: recognising the need for a new system; feasibility analysis; user requirements capture; choosing the optimum among the adaptation alternatives; planning the adaptation process; and implementation. The cases were given a list of users consisting of 1) senior managers; 2) construction project manager; 3) IT support; 5) end users; 6) Change agent/consultant, and were asked to identify who were actively involved in the different steps of the implementation procedure. The results obtained from the matrix question filled by the twelve companies are shown in Figure 3. The percentages indicate the number of companies stating that the user listed in that column is involved in the collaborative environment implementation step shown on the left of the figure.

Inspection of Figure 3 reveals that the involvement of end users was limited to user requirements capture. End users were involved mainly at the post-implementation training. It was found that IT support was involved in almost all stages. In companies 3, 4, and 6, the SMEs had not actually been involved in any of the decisions in the collaborative environment implementations other than agreeing on protocols and file formats to be used. Since the performance of the system depends on the users as well as the technical characteristics of the system, the needs of the user should be captured carefully. When the requirements of the users are met, they will work better through the system, improving the overall performance. The vitality of the timing of the implementation during the project was continually voiced by the interviewees. That is to say if the collaborative environment was introduced to a project team which was already formed from employees who had not used collaboration tool before, the adoption would be difficult. If it was introduced to this team when the project was already somewhere down the line, the result would most certainly be disastrous. Although no difference was found between contractors and consultancy SMEs in the collaborative environment implementation approaches apart from the transparency of the system affecting the contractors' trust, one of the biggest difficulties in the collaborative environment implementations on a construction project was to get the architecture SMEs to use the system. The architecture SME companies, and those referred to by other professions chose to do the work in the traditional way, without using the collaborative environment, and delegated the role of inputting all of their comments, sketches and mark ups to the system to a junior member of staff rather than doing it themselves. Although, managers are aware that maximising benefits can be attained when change is process led and not technology led, they constantly voiced concerns that achieving process led change is often more difficult in practice. Yet, where change was necessary in the interviewed companies, strategies were not mentioned neither achieved through business process reengineering or change management. This clearly indicates that processes that enable SMEs to agree a common vision and priorities for the collaboration manifested in a route map for how the project is going to proceed did

Figure 3. User involvement in collaboration environment implementation steps

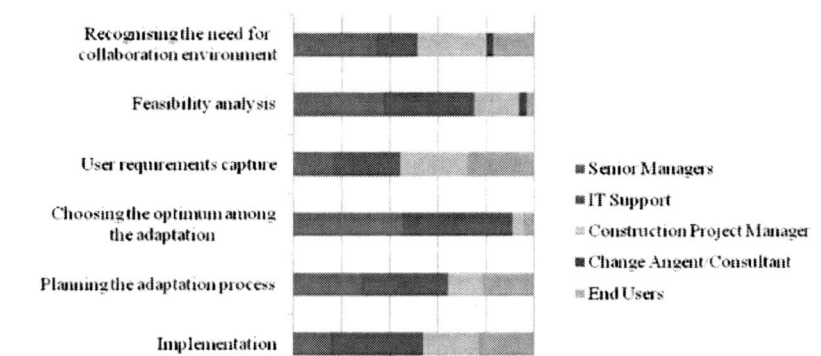

not exist, this must include suitable time for review of progress against vision and priorities.

Decision Making and Perception in Relation to Change

When the companies were asked how they implement a new collaborative environment, and how they handled the transformation occurring in the SME, the responses were mainly limited to training. The case study showed that there was no formal way of obtaining end user feedback throughout the implementation in any of the SMEs. In some of the companies, if the implementation included any testing or validation stage, the end-users were involved.

It was found that the companies did not have a formal method for capturing user requirements and only three companies focused on a sample of end users in order to get feedback after the collaborative environment was implemented. It was observed that the companies that do not involve the users in the requirements capture stage complained more about user resistance which shows excellent agreement with previously published results. As the project moves from one project to the next one with the same collaboration technology, if other conditions are similar, the success of the collaborative environment will increase. Standards that facilitate interoperability between different systems were voiced constantly by the interviewees.

It can be seen from Figure 3 that the IT support had a very active role in almost all stages of collaborative environment implementations. This situation is one of the results of too much focus on IT approach in the collaborative environment implementations. And the more IT people are involved in the implementations, the more focus on IT is observed. On the other hand, this situation can be used positively if the IT support can be influenced to consider sociocultural and organisational factors more in the implementation. If IT support considers the management concepts in the design and implementation of the collaborative environments, they may play a more important role in the adoption process. This clearly indicates that early user involvement is a critical factor for reducing user resistance to change.

Risk Management

The relatively high percentage of SMEs failure in providing the full benefits expected from the collaboration technologies could be attributed to incorrect or inadequate changes to processes, structure and their supporting IT systems. The growing awareness among large organisations of the benefits and importance of process improvements and technological developments related to collaboration places SMEs in a difficult position particularly in deciding how best to bring about business improvement with minimal risk. When the interviewees were asked what the key processes that ensure success of collaboration were, SMEs did not implement what they have in mind that technological solutions to business problems related to collaboration is not sufficient to improve the efficiency and effectiveness of work environments without carefully considering improvement to their current business processes. It is therefore highly important that the footsteps of a gradual and steady movement are followed to achieve this target which eventually reduces the risk of failure. This clearly indicates that understanding business processes becomes critical and must be carried out in advance of any IT investment decision in order to improve the opportunity to explore the full potential of technology in support of the overall business strategy.

Commitment and Buy-In

Commitment also needs to be managed. Most of the companies underlined the importance of top level commitment in motivating the employees to collaborate and to use the technology. If a change is to be introduced to an SME, the top level managers should first believe that this is necessary and act accordingly. They will have two roles: act as collaboration chiefs and manage the implementation, and ensure that the employees in the SME use the technology. It was clear from the study that in companies where implementation is let optional, the employees will continue to follow their old ways. They will not make an effort to get used to a new technology. Therefore, the senior manager must make it very clear that the new system has to be used in the SME. There should be someone from top level management putting the foot down to ensure that the collaboration technology is the only way of working in the SME. On the other hand, the study revealed that if this push was implemented in terms of coercion, then the employee resistance to change would not be observed but it might be transformed into a hidden rage which might create more problems in the future. In fact, pushing users to use the system should not be by coercion, but may involve making people know that they will have to face the consequences if they are not using the system. All the cases stressed the importance of the collaboration technology being used by all parties for the success of the whole project and they stated that it should be ensured by mutual agreement. Issuing the documents through the document controller, as in Companies 4, 6, 8, 11 and 12 could not be considered as a complete buy-in. as such, the managing directors need to pledge their support and ensure that the technology is implemented to ensure higher productivity and efficiency. Let alone the importance of the employees' satisfaction with current working conditions and interest and motivation for self-improvement and expanding their knowledge. Most companies lay emphasis on different methods of training to improve quality of performance.

This issue was voiced by some respondents concerned with employees' motivation and work satisfaction in collaborative environments implementation clearly indicating that employees must also be interested and motivated with IT which has to be clearly supported by top management. When the employees are satisfied, IT interest and attitude will change for the better.

Empowerment

When the interviewees were asked how collaborative environments can contribute to employee' empowerment, the results revealed that employee empowerment in the early stages reduces resistance to change, and allows for different perspectives to change. But, receiving project documents, making relevent decisions and responding in a timely and professional way through collaborative environments empowers team members, and thus teamwork enhances quality of work. The results revealed that collaborative environments designed to improve a business function for one culture, cannot be easily implemented in another culture. In companies 2 and 11 employees were not empowered to receive new project documents nor make relevant decisions regarding their particular profession. Any decision taken in response to a particular activity has to be endorsed by the management level. In these cases people felt powerless to respond to the other project participants in a timely and a professional way through collaborative environments. This has defeated the vision for which the environments were designed for. This clearly indicates that collaborative environments are unlikely to be successfully implemented unless SMEs change in line with the objectives of the environment.

Relationships and Trust

In all companies, work relationships were very formal and were dictated strictly in accordance with the contractual provisions provided. There was no room in any of the companies for relationship within the work environment to be different from what the arrangement required. The associated penalties that were imposed for non compliance and non performance meant that every SME within the project delivery team had to follow through its contractual obligations due to obvious financial implications. Although contractual agreement are highly important, teams that have come together within a short period of time can better function as a team when members are prepared to see beyond contractual limitations and obligations.

Promoting trust in the collaboration and the security of data were mentioned by all interviewees. A key person needs to be in charge, they provide leadership, leading hopefully to better performance of the team. It is important to ensure that the data on the system can be accessed only by the appropriate parties. Furthermore, the transparency of the system should be adjusted to a level that each SME's data is safe on the system. This was considered as a very important factor especially by contracting companies since they did not want to lose their bargaining power and their benefits from the claims for additional work carried out in similar work. It has been interpreted from the case study that the reservations will be less if the type of information to be shared and the extent of sharing are fixed at the beginning. The architecture company 11 thought that trust issue could be solved if the collaborative environment was implemented and led by a third party whereas the other architecture companies suggested that the collaborative environment should be led by the architecture companies since they are involved in the project from the very beginning and stay till the end. This clearly indicates that it is highly important to consider facilitating trust and sharing in any collaborative environments implementation.

Communication

The case study revealed that there are alternative conceptualisations of communication. Yet, it was listed as the root cause of most collaborative working failures. The interviewees, however, highlighted that collaborative environment solutions should reduce failure. All employees voiced concerns with regard to communication of change and improvements as essential in facilitating process and IT improvements. They stressed the importance of communication in successful implementation of change. Yet, it was considered by SMEs to be the most difficult aspect of implementation. In companies 9, and 10, efficient communication contributed significantly to the equal respect for all teams involved in the projects. This clearly indicates that it is important that communication takes place throughout the change process at all levels and for all individuals, and it should occur regularly between those in charge of the change initiatives and those affected by them. It is highly important that the SME structure, therefore, allows direct lines of communication across organisation boundaries. In fact, a flat and direct team structure ensures that everyone within the team has equal access to management and allows the whole team to act as a single unit.

Teamworking

Teamwork is viewed by all interviewees as the most important value in implementing collaborative environments confirming that teamwork brings about many advantages. The six key dimensions of teamwork were used as the major themes to determine the level of teamwork effectiveness, namely: single team focus and objectives; seamless operation without boundaries; unrestricted cross-sharing of information; creation of single and co-located team; equitable relationships, opportunities and respect for all; no blame culture. A teamwork matrix question focussed on the full, partial achievement and absence of the six main dimensions of team working within project organisations to determine the extent of integration within the team. The same data were used to determine the level of teamwork effectiveness within the various project delivery teams.

It was confirmed from the results that a fully teamworking is, therefore, expected to show a high level of teamwork effectiveness that will lead to an improvement in project delivery performance. Though, SMEs revealed that they struggle to achieve an agreed single focus and set of objectives, an espoused requirement for teamworking. Project teams in company 8, 11, and 12 were totally focussed on a single project objective. The members in these teams had the same focus and collectively worked as a group to achieve the objectives of the project. All the project teams were unable to operate seamlessly due to the continued operation of their members within their boundaries of organisational identity or affiliation. In Company 12 the various teams made significant efforts at collaborating with each other. In light of the previously published research, for effective teamworking to be fully realised, individual team identities must give way to a new single integrated team in which defined organisational boundaries do not exist. However, the extent to which seamless operation is a necessary condition appears questionable given that all the project teams were not able to form a new single team although they operated within a single office location. They remained as individual sub teams within their confined work spaces but collocated with others. They worked towards achieving a single team focus and common project objectives but did not operate seamlessly. The various members within the project team continued to work disjointly within their individual organisations. They were not able to form a new single team that was co-located. This is indicative of the SMEs struggling to overcome work attitudes even when they are negatively impacting on its performance.

Legal and Contractual

Lack of agreement between teams was found as the main failure problem in this dimension. The cases stressed the importance of the collaboration tool being used by all parties for the success of the whole project and they stated that as well as mutual agreement, it should also be ensured by contractual terms. One of the biggest difficulties in the collaborative environment implementations on a construction project was to get the architecture SMEs to use the system. According to the observation from the interviews, the architecture SMEs would use a collaborative environment only if the contractual agreement with the client obliged all team members to use the collaboration system. The importance of contract terms regarding the collaborative environment used for external communication was particularly emphasised by the companies. It was continually voiced that contract binding for all companies participating in the project to make sure there are consistent procedures for the use of the systems are vital. This clearly indicates that the legal context and infrastructure governing the relation among the various parties and addressing the challenges such as information ownership, liability for the loss of information and/or breakdown of the system and insolvency of the technology providers is not properly addressed or clearly stipulated in the conditions of the contracts governing the relations of the case study.

The case study revealed that the legal infrastructure and forms of contracts which foster the collaboration environments implementation were the main reasons behind miscommunication problems. The legal infrastructure and available forms of contract do not fairly distribute elements of risks among various parties involved which further aggravates the communication and authority situation. On the other hand, all communications are logged in a secure environment which will facilitate future tracing back of documents. This means that traditional mistakes generated from someone working on an old document or drawing are eliminated or at the very least minimised. More crucially such an environment reduces the opportunity for mistakes leading to disputes which is the biggest cause of waste and inefficiency in the construction industry. A right balance between the technology and professional liability is the issue to building trust. Another issue raised by some respondents is an initial workshop involving the project team organised to get the project team members to agree on their project protocol-similar to the Project Information Exchange protocol. It is therefore necessary for the collaborating SMEs to agree on the common formats, types and conventions for the information exchange before the collaborative environment is set up to provide consistency and avoid possible confusion.

RECOMMENDATIONS

Based on the factors that have been found to give impetus to the challenges associated with collaborative environments, it is recommeded that these are addressed in SMEs to determine the success of collaboration. The qualitative evidence revealed five main challenges associated with implementing collaborative environments initiatives, namely: lack of top management support; the lack of a provision for appropriate training; the creation of an appropriate culture; the need to adopt appropriate processes; the adoption of appropriate technologies. The lack of top management support is the main challenge associated with implementing collaborative environments because of a lack of awareness of collaborative environments benefits, lack of vision and strategy; and the lack of structure for collaborative environments initiatives. The provision of appropriate training was revealed as the second main challenge. This is due to lack of time, budget, and difficulties associated with training. Empowering people to collaborative working is equally important.

Employees' commitment is concerned with ensuring that managers not only encourage collaboration, but also show that they actively support and participate in collaboration initiatives. The adoption of appropriate processes for collaborative working was revealed as the fourth main challenge associated with implementing collaborative environments. There are many challenges associated with collaborative environments initiatives in the construction industry such as the need for an appropriate culture, the requirement for changing the attitude of employees, the importance of sufficient budget, and the identification of the most appreciated collaboration tools. Collaborative environments is an integrated and complex social process, which has management, training, contractual obligations, people, technology, communication, and organisational structure at its core.

The qualitative data analysed revealed three factors associated with collaborative environments that contribute to improved competitiveness. These are improved efficiency, improved productivity and increased profitability. Since collaborative environments generally fail to achieve the full benefits expected and the reason for this failure is not technical but related to how it is implemented with regards to organisational, legal, and socio-cultural related factors, these favtors ought to be given due estimation. However, the most interesting aspect is that the companies did not have any defined criterion to measure the success of the collaborative environment implementations and mostly carried out a perceptual analysis of whether their employees worked better than previously and whether they were more efficient or more useful than previously. As such, SMEs need a different structured procedure for collaboration implementation and management, together with a detailed change management approach to control all the factors affecting the success of collaboration environments.

REFERENCES

Anumba, C. J., Baugh, C., & Khalfan, M. A. (2002). Organsiational structures to support concurrent engineering in construction. *Industrial Management & Data Systems, 102*(5), 260–270. doi:10.1108/02635570210428294

Arendt, C., Landis, R., & Meister, T. (1995). *The human side of change – Part 4* (pp. 22–27). IIE Solutions.

Baiden, B. K., Price, A. F., & Dainty, A. R. (2003). Looking beyond processess: Human factors in team integration. In Greenwood, D. J. (Ed.), *ARCOM*. Brighton.

Baldwin, A. N., Thorpe, A., & Carter, C. (1999). The use of electronic information exchange on construction alliance projects. *Automation in Construction, 8,* 651–662. doi:10.1016/S0926-5805(98)00110-1

Barbour. (2002). *The Barbour Report 2002: Exploring the Web as an information tool: A practical guide*/ Windsor, Canada: Barbour Index.

Barbour. (2003). *The Barbour Report 2003: Influencing clients: The importance of the client in product selection.* Windsor, Canada: Barbour Index.

Bromley, S., Worthington, J., & Robinson, C. (2003). *The impact of integrated teams on the design process.* London, UK: Construction Productivity Network.

Carter, C., White, E., Hassan, T., Shelbourn, M., & Baldwin, A. (2002). Legal issues of collaborative electronic working in construction. *Proceedings of the Institution of Civil Engineers, Civil Engineering, Special Issue Two: Information Technology - The Key to Collaboration,* (pp. 10-16).

Cornick, T., & Mather, J. (1999). *Construction project teams: Making them work profitably.* London, UK: Thomas Telford Department for Environment, Transport and the Regions. (2000). *Construction statistics annual.* London.

Department of Trade and Industry. (2004). *Business in the information age: The international benchmarking study.* London, UK: Author.

E-Business W@tch. (2006). *ICT and e-business in the construction industry.* Sector Report No. 7, European Commission.

Erdogan, B., Anumba, C., Bouchlaghem, D., & Nielsen, Y. (2010). An innovative integrated framework towards effective collaboration environments in construction. *International Journal of Technology Management, 50*(2), 139–168. doi:10.1504/IJTM.2010.032270

Feeny, D., & Willcocks, L. (1998). Core IS capabilities for exploiting Information Technology. *Sloan Management Review,* 9–21.

Hammer, M., & Stanton, S. (1995). *The re-engineering revolution.* New York, NY: Harper Collins.

Harris, P. R., & Harris, K. G. (1996). Managing effectively through teams. *Team Performance Management, 2*(3), 22–36. doi:10.1108/13527599610126247

Hassan, T., Shelbourn, M., & Carter, C. (2008). *Collaboration in construction: Legal and contractual issues in ICT application.*

Hegazy, T., Zaneldin, E., & Grierson, D. (2001). Improving design coordination for building projects. I: Information model. *Journal of Construction Engineering and Management, 127*(4), 322–329. doi:10.1061/(ASCE)0733-9364(2001)127:4(322)

Holt, G. D., Love, P. E. D., & Nesan, L. J. (2000). Employee empowerment in construction: An implementation model for process improvement. *Team Performance Management, 6*(3/4), 47–51. doi:10.1108/13527590010343007

Kitchen, P. J., & Daly, F. (2002). Internal communication during change management. *Corporate Communications, 7*(1), 46–53. doi:10.1108/13563280210416035

McAdam, R., & Galloway, A. (2005). Enterprise resource planning and organisational innovation: A management perspective. *Industrial Management & Data Systems, 105*(3), 280–290. doi:10.1108/02635570510590110

Mumford, E. (1995). Creative chaos or constructive change: Business process-re-engineering versus socio-technical design. In Burke, G., & Peppard, J. (Eds.), *Examining business process re-engineering: Current perspectives and research directions* (pp. 192–216). New York, NY: Kogan Page.

Murdoch and Hughes. (2008). *Construction contracts: Law and management*. Spon Press.

Pena-Mora, F., & Dwivedi, G. H. (2002). Multiple device collaborative and real time analysis system for process management in civil engineering. *Journal of Computing in Civil Engineering, 16*(1), 23–37. doi:10.1061/(ASCE)0887-3801(2002)16:1(23)

PIECC. (2006). Planning and implementation of effective collaboration within construction/ Retrieved July 10, 2009, from http://piecc.lboro.ac.uk/

Proctor, T., & Doukakis, I. (2003). Change management: The role of internal communication and employee development. *Corporate Communications: An International Journal, 8*(4), 268–277. doi:10.1108/13563280310506430

Rezgui, Y. (2007). Exploring virtual team-working effectiveness in the construction sector. *Interacting with Computers, 19*(1), 96–112. doi:10.1016/j.intcom.2006.07.002

Rezgui, Y., Wilson, I., Olphert, W., & Damodaran, L. (2005). Socio-organizational issues. In Camarinha-Matos, L. M., Afsarmanesh, H., & Ollus, M. (Eds.), *Virtual organizations systems and practices*. New York, NY: Springer. doi:10.1007/0-387-23757-7_13

Robson, W. (1997). *Strategic management and Information Systems*. London, UK: Pitman Publishing.

Roshani, D., & Tizani, W. (2005). *Integrated IFC based collaborative building design using internet technology*. The Tenth International Conference on Civil, Structural and Environmental Engineering Computing, Rome, Italy.

Ruikar, K., Anumba, C. J., & Carillo, P. M. (2005). End-user perspectives on use of project extranets in construction organsiations. *Engineering, Construction, and Architectural Management, 12*(3), 222–235. doi:10.1108/09699980510600099

Rye, C. (1996). *Change management action kit*. London, UK: Kogan Page.

Small Business Service. (2004). *Statistics*. Retreived April 2, 2009, from http://www.sbs.gov.uk/ default.php? page=/statistics/dcfault.php

Thorpe, T., & Mead, S. (2001). Project-specific web sites: Friend or foe? *Journal of Construction Engineering and Management, 127*(5), 406–413. doi:10.1061/(ASCE)0733-9364(2001)127:5(406)

Underwood, J., & Alshawi, M. (2000). Forecasting building element maintenance within an integrated construction environment. *Automation in Construction, 9*(2), 169–184. doi:10.1016/S0926-5805(99)00003-5

Weippert, A., & Kajewski, S. L. (2002). Internet-based information and communication systems on remote construction projects: A case study analysis. *Construction Innovation, 2*(2), 103–116.

Wilkinson, P. (2005). *Construction collaboration technologies: The extranet evolution*. London, UK: Spon Press.

KEY TERMS AND DEFINITIONS

Collaborative Environment: An environment created using different collaboration technologies where all organisations involved in the construction project can communicate, exchange data and information, and carry out joint activities in order to realise the successful completion of this project.

Legal and Contractual: Legal implications of using collaboration technologies, and legal protection inherent in contractual provisions.

Organisational: Of, relating to, or produced by an organisation.

Small and Medium Enterprises: Enterprises employing less than 250 employees.

SocioCultural: Culture as a social context in which collaboration occurs.

Chapter 10

E–Readiness in IT/ IS Implementation:
A Benchmarking Analysis Based on 100 Case Studies

Ayman Altameem
University of Bradford, UK

Mohamed Zairi
University of Bradford, UK

EXECUTIVE SUMMARY

The value that information technology (IT) can bring to organizations is clear, and few will dispute its potential. However, the literature shows that worldwide, many of the organizations adopting IT fail to achieve the desired results and that sometimes, the cost of failure can far exceed the expected benefits. There are several reasons why IT projects fail to deliver. The study which is described in this chapter was conducted for the following reasons, all related to the level of readiness during IT Implementation: 1) Many organizations do not rely on IT just for reasons of efficiency, lowering cost, or for even improving productivity. Rather, their long-term existence depends on its successful use. 2) IT projects are often considered as high-risk projects. It is reported that as little as only 16% of IT projects were considered successful, mainly due to poor adoption practices. 3) Increased IT investments have not resulted in adequate returns. The return on investment has been consistently discouraging in terms of the gap between expectation and realization. 4) Studying the factors which lead to IT adoption is an important issue. Completing IT projects

DOI: 10.4018/978-1-61350-311-9.ch010

successfully the first time requires the identification and understanding of all the critical success factors.

This study is an attempt to bridge the gap in the existing literature by exploring the critical factors that affect IT adoption through a comprehensive benchmarking analysis, using secondary cases. The IT adoption in 100 organizations indicated in the literature, were scrutinized in all the cases analyzed in order to arrive at the most critical factors affecting IT adoption, as well as their degree of criticality. The study identifies twenty-four critical factors that must be carefully considered in IT adoption to attain successful outcomes.

INTRODUCTION

In a global organizational context, IT is no longer mainly confined to backroom operations (Dehning et al., 2005), it has risen beyond its traditional support role and now plays a central role in formulating business strategies (Chan, 1997). As Bob Martin, CEO of Wal-Mart's International Division, says, "At Wal-Mart and at many other companies, technology has become integrated with every aspect of the business" (Dehning et al., 2005). Nowadays, control and co-ordination would be impossible without IT.

The growth of IT has rapidly changed the features of business over the past decade. Dewett and Jones (2001) state that "the availability and use of information systems and technologies has grown almost to the point of being commodity-like in nature, becoming nearly as ubiquitous as labour". Various reports that this new revolution in the use of IT shows how significantly this technology is changing many aspects of today's business activities and numerous researchers believe these changes will accelerate over the coming years.

In order to maximize the benefits from investing in IT, organizations must understand how to manage their adoption process, the critical factors affecting their adoption and the benefits and obstacles which may emerge from IT adoption.

Since the mid-20th century, the world has been undergoing an IT revolution; as a result IT has extended itself into every corner of the world and every aspect of our individual lives. The impact of IT is reflected in the growth of the world economy, in the operation of civilized societies and organizations in general, as well as the lives of individual people (Awallmah, 2002).

However, IT adoption is challenging and risky (Kumar, 2002; Zee, 2002; Milis and Mercken, 2004; Rodriguez-Repiso et al., 2007). Thus, it is necessary for organizations to understand the critical factors of IT adoption, and studying such factors is an important issue (Swanson and Wang, 2005; Tarafdar and Vaidya, 2006; Bruque and Moyano, 2007). Brandon (2006) states that: "Completing IT projects successfully the first time requires the identification and understanding of all the critical success factors". Thus, a clear study of the essential elements in the working field environment through secondary case studies will supplement the literature for better understanding of the critical factors which may affect organizations' success in adopting IT.

Since IT still has many problems or shortcomings associated with its adoption, it is vital to profit from organizational experience. Thus, it is crucial to look at what others have done and consider their feedback, mistakes, results and overall approach to IT project adoption. Consequently, this chapter provides a comprehensive analysis of 100 secondary case studies of IT adoption presented in the literature and reviewed by the researcher, in order to arrive at the most critical factors of IT adoption.

FAILURES IN IT/IS IMPLEMENTATION: WHAT ARE THE READINESS ISSUES?

More or less everything in the world has both successes and failures. Throughout the world, organizations are investing enormous resources in IT in order to increase their market share (Lauria and Duchessi, 2007) and there is evidence of its positive impact in various organizations (O'Brien and Al-Biqami, 1999; Osei-Brysona and Kob, 2004). However, adopting IT is often expensive and risky (Kumar, 2002; Zee, 2002; Milis and Mercken, 2004; Rodriguez-Repiso et al., 2007) and track records show discouraging outcomes.

Instead of gradually dropping over time, the IT project failure rate remains high (Booty, 1998). Keil, Mann and Rai (2000) note that 30-40% of all projects are less successful than promised and at least one in four projects (25%) ends in failure. Furthermore, a 2003 survey conducted by Oxford University and *Computer Weekly* in the UK shows that more than 80% of IT projects were considered a failure (Sauer and Cuthbertson, 2003).

In addition, Standish Group International conducted a study in 1995, which revealed data from thousands of IT projects, showing that a total of 46% involved not only costs higher than the original estimates, but were completed behind schedule, while 28% were terminated before completion. Only 26% of projects ended on time, and within the estimated budget (Saleh and Alshawi, 2005). Gartner Group (2001) reported that, on average, organizations spent US$1 million each year on unsuccessful projects, on top of wasted professional resources which cannot be easily quantified, and that about 40% of IT projects failed.

Systems may not work or may fail to deliver and serve the purpose. Some IT failures involve deadlines being over-run, out-of-control maintenance, exceeded budgets and frequently the quality of the new system being far below the expected standard when the project was undertaken. The return on investment has been continually discouraging in terms of the gap between expectation and realization. According to Willcocks and Lester (1994), few IT projects realize any positive benefits, while Clegg et al. (1997) found that 80-90% of IT investments do not meet their performance objectives. The UK government scrapped a computer system of about £140 million to save about £60 million through simplifying the system (Wilson, 2006). Accordingly, the £140 million benefits processing reimbursement programme ran into a series of technical difficulties. Similarly, there are studies that point out that the increased IT investments in developing countries have not resulted in adequate returns (Ranganathan and Kannabiran, 2004). This confirms that the aggregate return on investment is discouraging. Lyytinen and Hirschheim (1987) define four major categories of IT failures, as shown in Table 1.

Table 1. Major categories of IS failures

Category	Description
1. Correspondence failure	When the system's design objectives are not met, the IS is considered a failure. It is generally believed that design goals and requirements can be specified clearly in advance, and that their achievement can be accurately measured. Performance measures mainly based on cost-benefit analysis are employed for managerial control over the system's implementation. Correspondence failure, goal seeking in outlook, tends to neglect the fact that users may not necessarily accept systems meeting design objectives and specifications.
2. Process failure	Process failure occurs when an IS cannot be developed within an allocated budget, and/or time schedule. There are two likely outcomes of process failure. Firstly, an outright failure occurs when no workable system can be produced. Secondly, a more common outcome is when an IS is developed with massive overspending in both cost and time, thus negating global benefits of the system. This is a project level failure attributed to unsatisfactory project management performance.
3. Interaction failure	The level of end-user usage of the IS is suggested as a surrogate in IS performance measurement. Some related measures of IS usage include user attitudes and user satisfaction, the amount of data transferred or the frequency of use. However, heavy usage does not necessarily mean high user satisfaction and improved task performance, and there is little empirical evidence supporting such a claim. Heavy systems usage might be a result of legal compulsion, persuasion, or that there are simply no other alternatives besides using the system.
4. Expectation failure	The notion of expectation failure views IS failure as the inability of a system to meet its stakeholders' requirements, expectations or values. Failure, therefore, does not only involve the system's inability to meet design (technical) specifications. Expectation failure is perceived as the difference between the actual and desired situation for the members of a particular stakeholder group. Unlike the other three notions, IS failure is considered holistically in this case, as the views of different stakeholders are taken into account.

Source: Adapted from Lyytinen and Hirschheim (1987)

Overall, IT investment is at an all-time high and is increasing significantly. However, according to Avgerou and Cornford (1998), organizations have learned that managing the technologies themselves is a risky and complex job which causes stress to its own experts. Thus, Checkland and Holwell (1998) consider that the study of IS remains a crucial but confused field.

BACKGROUND TO CASE STUDIES

The 100 case studies contributing to the present analysis have been gathered from different sources, such as articles, reports and IT vendors' case studies published online through their websites. The survey spanned different countries and various types of sectors, such as manufacturing, government, education, transportation, finance and insurance, retail, healthcare, technology and others. Table 2 gives a full list of these organizations. For ease of reference, the table provides details of the particular focus, location and sources of every case study. The choice of cases for analysis was based on the availability of information.

All factors were identified and classified based on the framework of IT adoption (Figure 1). This conceptual model consists of the 24 factors and is shown below.

The major reason for using this is that it is based on an extensive literature review. The researchers used a content analysis approach; the number of times authors mentioned the factor were coded to analyze the degree of criticality. The main objective of analyzing these cases was to extract the most critical factors required

Table 2. 100 organizations used as secondary case studies

No.	Organization Name	Industry	Country	Source
1	7-Eleven, Inc.	Retail	USA	EDS (2007)
2	Airport Authority Hong Kong	Transportation	Hong Kong	EDS (2008)
3	Al-Bahar	Retail	Arabian Gulf	Farabi case study (2002)
4	Aloha Airlines	Transportation	USA	IBM case studies (2008)
5	Amazon	Retail	USA	Timmers (1999)
6	American Airlines	Transportation	USA	EDS (2007)
7	American University of Beirut	Education	Lebanon Middle East	Dell case studies (2005)
8	An Garda Síochána	Government	Ireland	Fujitsu (2000)
9	Auckland City Council	Government	New Zealand	Microsoft case study (2008)

continued on following page

Table 2. Continued

No.	Organization Name	Industry	Country	Source
10	Australian Taxation Office (ATO)	Government	Australia	EDS (2007)
11	Austrian National Library	Government	Austria	Dell case studies (2006)
12	Avista Corporation	Energy	USA	EDS (2007)
13	Banque Hervet	Finance/ Insurance	France	EDS (2008)
14	Barbecue Plaza	Retail	Thailand	IBM case studies (2008)
15	Barnes and Nobel	Retail	USA	Al-Mashari (2002)
16	Baylor	Healthcare	USA	EDS (2007)
17	Bedfordshire Police	Government	UK	Dell case studies (2005)
18	Beijing East Express Logistics Co., Ltd.	Transportation	China	Fujitsu (2008)
19	BHP Steel	Manufacturing	Multi-national	Caroline and Paula (2000)
20	Blue Nile	Retail	USA	Cisco case (2001)
21	BMI	Transportation	UK	Fujitsu (2001)
22	BP	Energy	Canada	EDS (2007)
23	British Airways	Transportation	UK	IBM (2008)
24	Charle Co., Ltd	Retail	Japan	Fujitsu (2008)
25	Charleston County	Government	USA	Dell case studies (2003)
26	Chinatrust	Finance/ Insurance	Taiwan	FNS case (2001)
27	City of Anaheim	Government	USA	EDS (2007)
28	City of Los Angeles	Government	USA	IBM case studies (2007)
29	Coles Group Limited	Retail	Australia	EDS (2007)
30	Coors Brewing Company	Retail	USA	EDS (2007)
31	County Materials Corporation	Manufacturing	USA	Microsoft case studies (2008)
32	Del Monte Foods	Manufacturing	USA	Microsoft case studies (2008)
33	Dell	Technology	USA	Dell case studies (2007)
34	Dexia Bank	Finance/ Insurance	Belgium	EDS (2007)
35	Dublin Airport Authority	Transportation	Ireland	Fujitsu (2008)
36	eBay	Auction	Multi-national	David (2001)
37	Egg	Finance/ Insurance	UK	Clarke and Doherty (2004)
38	EMS Group	Manufacturing	USA	Microsoft case studies (2008)

continued on following page

Table 2. Continued

No.	Organization Name	Industry	Country	Source
39	Environmental Protection Agency (EPA)	Government	USA	Microsoft case studies (2008)
40	FedEx	Air Courier	Multi-national	Timmers (1999)
41	Ford Motor	Manufacturing	Australia	Ratnasingam (2001)
42	Fortis	Finance/ Insurance	UK	Baets (1996)
43	Freightliner LLC	Manufacturing	USA	Microsoft case studies (2008)
44	Garanti Bank	Finance/ Insurance	Turkey	IBM (2001c)
45	Global Financial Services Leader	Finance/ Insurance	Multi-national	EDS (2008)
46	Honeywell	Manufacturing	USA	Microsoft case studies (2008)
47	IBM	Technology	USA	IBM case studies (2007)
48	Intel	Semiconductors	USA	Microsoft case studies (2008)
49	ITT industries	Manufacturing	Germany	Dell case studies 2006)
50	Jardine Air Terminal Services (JATS)	Transportation	Hong Kong	Fujitsu (2008)
51	JSC RusHydro	Government	Russian Federation	IBM case studies (2008)
52	KeyCorp	Finance/ Insurance	Multi-national	AIT (2001)
53	Kings College	Education	UK	Dell case studies (2006)
54	Komatsu Europe International (KEISA)	Manufacturing	Europe	EDS (2007)
55	Korean Association of Bookstores	Retail	Korea	Fujitsu (2008)
56	Linksys	Network Hardware Manufacturing	USA	IBM case studies (2008)
57	Lloyds TSB	Finance/ Insurance	UK	AIT (2001)
58	Meridian Bancorp	Finance/ Insurance	USA	Rajiv and Robert (1998)
59	MERSCORP, Inc.	Finance/ Insurance	USA	EDS (2008)
60	Metro Inc.	Retail	Canada	IBM case studies (2008)
61	Ministry of Economics & Labour	Government	Austria	EDS (2008)

continued on following page

Table 2. Continued

No.	Organization Name	Industry	Country	Source
62	Ministry of Transportation	Government	Italy	EDS (2007)
63	Mitsui Oil and Gas	Energy & Utilities	Japan	Fujitsu (2008)
64	Monsanto	Manufacturing	Belgium	Dell case studies (2006)
65	Mövenpick	Retail	Switzer-land	EDS (2007)
66	NAV	Transportation	Canada	IBM case studies (2007)
67	NCCC	Finance/ Insurance	Taiwan	EDS (2008)
68	NedTrain	Transportation	Nether-lands	EDS (2007)
69	OCBC Bank	Finance/ Insurance	Singapore	Microsoft case studies (2008)
70	Panasonic Europe	Electronics	Germany	IBM case studies (2008)
71	Pennsylvania Office of the Attorney General	Government	USA	Microsoft case studies (2007)
72	Philips Semiconductor	Technology	Germany	IBM case studies (2006)
73	Pondicherry	Government	India	Bowonder et al. (2000)
74	Province of Genoa	Government	Italy	Microsoft case studies (2008)
75	RailAmerica	Transportation	USA	EDS (2007)
76	SARA	Technology	Netherlands	Dell case studies (2007)
77	SaskEnergy	Government	Canada	EDS (2007)
78	Sharp Electronics Corporation	Electronics	USA	IBM case studies (2008)
79	Silicon Operations	Manufacturing	USA	Microsoft case studies (2008)
80	Statistics New Zealand	Government	New Zealand	Microsoft case studies (2008)
81	Syracuse Police Department	Government	USA	IBM case studies (2008)
82	TASER	Manufacturing	USA	Microsoft case studies (2008)
83	Technogym	Manufacturing	Italy	Microsoft case studies (2008)
84	The City of Amersfoort	Government	Netherlands	Dell case studies (2006)
85	The Highland Council	Government	UK	Fujitsu (2001)
86	The Royal Air Force	Government	UK	Fujitsu (2008)
87	Transtec	Manufacturing	Multi-national	Claudia and Stefan (2001)
88	Trimm Technologies	Manufacturing	USA	Fujitsu (2008)

continued on following page

Table 2. Continued

No.	Organization Name	Industry	Country	Source
89	TriView	Finance/ Insurance	USA	BNQP (2001)
90	University of Michigan	Education	USA	Dell case studies (2005)
91	University of Texas	Education	USA	Dell case studies (2006)
92	US agency	Government	USA	IBM case studies (2008)
93	Veneto Region	Government	Italy	Microsoft case studies (2008)
94	Welch Allyn	Technology	USA	IBM case studies (2007)
95	West Marine	Manufacturing	USA	IBM case studies (2001)
96	Westminster City Council	Government	UK	Microsoft case studies (2008)
97	Whirlpool	Technology	USA	IBM case studies (2006)
98	William Beaumont Hospital	Healthcare	USA	Dell case studies (2006)
99	Wilton	Retail	USA	EDS (2007)
100	Woolwich	Finance/ Insurance	UK	AIT (2001)

Figure 1. IT adoption framework

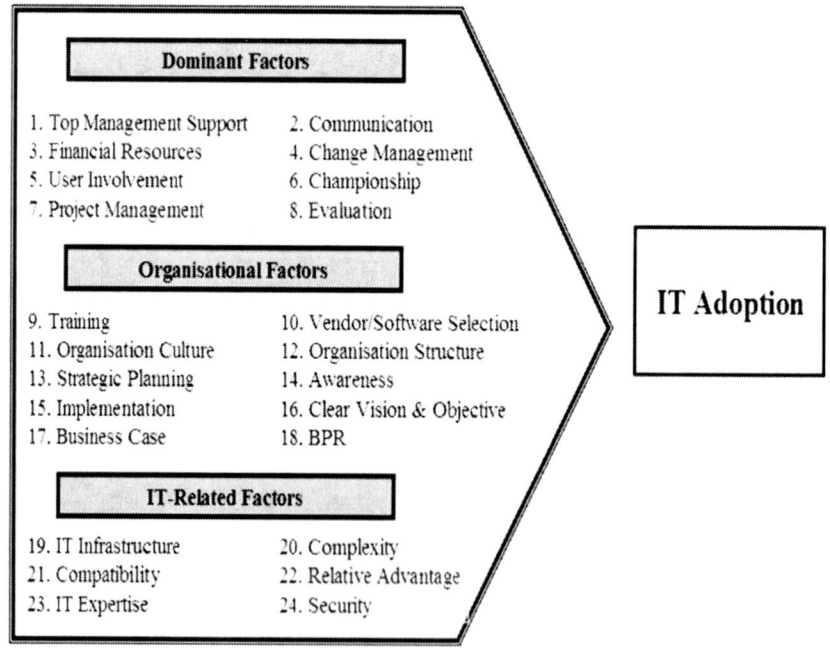

Table 3. Critical factors used to analyze selected cases

No.	Dimension
	Dominant Factors
1.	Top management support
2.	Change management
3.	Championship
4.	Financial resources
5.	Communication
6.	Evaluation
7.	User involvement
8.	Project management
	Organizational Factors
9.	Training
10.	Organization structure
11.	Strategic planning
12.	Business case
13.	Business process re-engineering (BPR)
14.	Organization culture
15.	Awareness
16.	Implementation
17.	Clear vision & objectives
18.	Vendor/software selection
	IT Factors
19.	IT infrastructure
20.	Relative advantage
21.	Complexity
22.	IT expertise
23.	Security
24.	Compatibility

for successful adoption of IT. Table 3 illustrates all of the 24 critical factors derived from the extensive review of the literature.

Figure 2. Organization regions

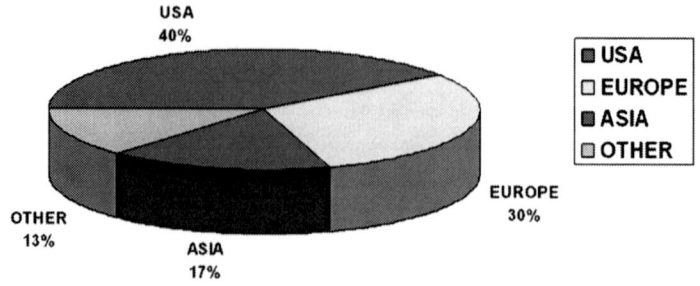

ANALYSIS AND RESULTS

General Results on IT Adoption

The 100 case studies analyzed originated from organizations from various sectors, based in a variety of countries. Figure 2 reveals the regions of these organizations. They have been classified by continent (Asia, USA, Europe and other). It can be observed that most of these organizations are from the USA (40%), whereas Europe came in second place (30%), followed by Asia and other regions (17% and 13%, respectively).

Figure 3 reveals that IT has made inroads into all sectors. The result shows that the IT project is more appealing to the government sector with 24%. However, there are about 17% of organizations from the manufacturing sector, and 15% from the Finance/ Insurance sector, closely followed by the retail sector with 14%. On the

Figure 3. Organization sectors

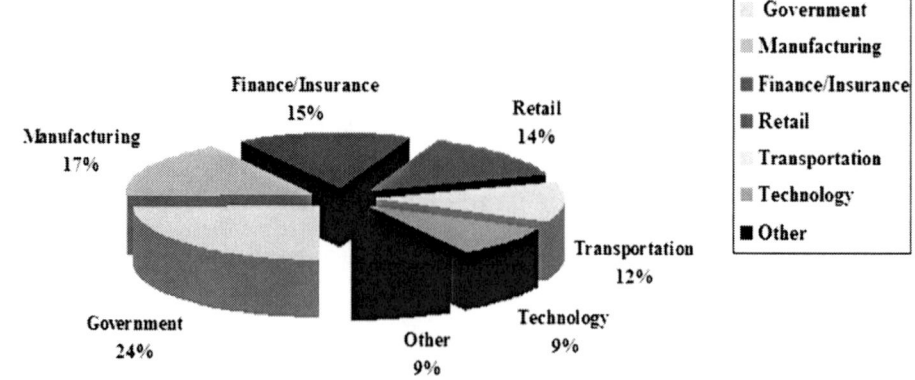

other hand, the figure shows that transportation, technology and others sectors are only 12%, 9% and 9%, respectively.

Critical Factors of IT Adoption

The analysis of the 100 case studies revealed many critical factors that influence the adoption of an IT project. Due to space limitations, Table 4 has been compiled to enable the user to view the range of factors within each organization. In addition, for simplicity, each dimension has been analyzed separately, as follows:

- Dominant factors
- Organizational factors
- IT-related factors

The percentage of the degree of criticality of each factor is presented in Figure 4. The figure shows that the most critical factor in successful IT project adoption, which stands out over all others, is top management support and commitment. Not surprisingly, the adoption of a new innovation is doubtful to succeed if it does not have support and commitment from the top management. In fact, for any project, especially a large project such as IT adoption, support and commitment of top management is indispensable to attain success (Hwang et al., 2004; Soliman and Janz, 2004; Kim and Lee, 2007). Slevin and Pinto (1986) state: "Early in a new project's life, no single factor is as predictive of its success as the support of top management". The following sections analyze all other critical factors of IT adoption.

Dominant Factors

As Figure 5 shows, the most critical factor in IT adoption in this dimension, which stands out over all others, is top management support and commitment, cited in 75% of case studies. This factor was the most significant in all dimensions, and was supported by many researchers and practitioners (Kwon and Zmud, 1987; Damanpour, 1991; Premkumar and Ramamurthy, 1995; Rogers, 1995; McFadden, 1996; Grover 1998; Premkumar and Roberts, 1999; Thong, 1999; Turban et al., 2000; Hwang et al., 2004; Soliman and Janz, 2004; Kim and Lee, 2007). Next in terms of high criticality in the dominant factors is communication, with 57%. Communication, through the life cycle of the project, is seen as an important aspect to enhance the chances of project success. In addition, Figure 6 shows that championship is found to be the third most critical factor with 53%, while project management, financial resources and evaluation come next with 50%, 38% and 22%, respectively.

Table 4. Critical factors in secondary case studies

No.	Name of organization	Top Management Support & Commitment	Communication	Evaluation	Championship	Change Management	User Involvement	Financial Resources	Project Management	Business Case	Awareness	Training	Organization Structure	Organization Culture	Clear Vision & Objective	Implementation	Strategic Planning	BPR	Vendor/Software Selection	IT Infrastructure	Compatibility	Relative Advantages	IT Expertise	Complexity	Security
1	7-Eleven, Inc.	X	X		X				X						X	X	X	X		X	X	X	X	X	
2	Airport Authority Hong Kong			X					X		X				X	X	X			X	X	X	X	X	
3	Al-Bahar	X			X			X		X						X	X		X	X					
4	Aloha Airlines	X			X			X										X	X	X					
5	Amazon	X	X																						
6	American Airlines	X	X	X		X			X	X					X	X	X	X		X	X	X		X	X
7	American University of Beirut	X		X	X				X	X				X	X	X	X		X	X	X	X	X		
8	An Garda Síochána	X	X	X	X	X		X	X	X		X			X	X	X	X		X	X	X	X		X
9	Auckland City Council	X	X	X	X	X		X	X	X	X	X			X	X	X	X	X	X	X	X	X	X	X
10	Australian Taxation Office (ATO)	X	X			X			X	X						X			X	X			X		

continued on following page

Table 4. Continued

No.	Name of organization	Top Management Support & Commitment	Communication	Evaluation	Championship	Change Management	User Involvement	Financial Resources	Project Management	Business Case	Awareness	Training	Organization Structure	Organization Culture	Clear Vision & Objective	Implementation	Strategic Planning	BPR	Vendor/Software Selection	IT Infrastructure	Compatibility	Relative Advantages	IT Expertise	Complexity	Security
11	Austrian National Library	×			×							×				×			×	×	×				
12	Avista Corporation		×					×							×		×	×		×					
13	Banque Hervet	×			×				×	×	×					×	×	×	×	×			×		
14	Barbecue Plaza	×			×			×	×	×					×	×	×	×	×	×	×	×	×	×	×
15	Barnes & Nobel	×													×		×		×		×	×			×
16	Baylor	×						×					×			×				×				×	
17	Bedfordshire Police	×	×		×					×	×	×			×	×	×	×	×	×	×	×			
18	Beijing East Express Logistics Co., Ltd.									×	×	×		×	×		×		×	×	×	×			
19	BHP Steel	×	×	×							×	×			×		×								
20	Blue Nile	×	×		×			×		×					×		×	×	×	×	×	×		×	×
21	BMI	×	×		×		×		×					×	×	×	×		×	×	×	×	×	×	

continued on following page

235

Table 4. Continued

continued on following page

No.	Name of organization	Top Management Support & Commitment	Communication	Evaluation	Championship	Change Management	User Involvement	Financial Resources	Project Management	Business Case	Awareness	Training	Organization Structure	Organization Culture	Clear Vision & Objective	Implementation	Strategic Planning	BPR	Vendor/Software Selection	IT Infrastructure	Compatibility	Relative Advantages	IT Expertise	Complexity	Security
		Dominant factors / Organizational factors / IT-related factors													IT critical factors										
22	BP	X						X	X						X	X	X	X			X	X	X	X	X
23	British Airways	X	X		X				X	X					X	X	X	X	X	X	X	X			
24	CHARLE Co., Limited	X	X		X				X	X					X	X	X		X	X			X	X	
25	Charleston County			X	X		X			X						X	X		X	X	X	X	X		
26	Chinatrust	X			X	X			X	X							X								
27	City of Anaheim	X	X					X							X	X				X	X	X	X	X	X
28	City of Los Angeles									X									X	X	X	X			
29	Coles Group Limited	X	X	X	X			X	X	X					X	X	X	X	X	X	X	X		X	
30	Coors Brewing Company											X			X	X	X		X	X			X		
31	County Materials Corporation	X	X							X	X				X	X	X		X	X		X		X	

236

Table 4. Continued

No.	Name of organization	Top Management Support & Commitment	Communication	Evaluation	Championship	Change Management	User Involvement	Financial Resources	Project Management	Business Case	Awareness	Training	Organization Structure	Organization Culture	Clear Vision & Objective	Implementation	Strategic Planning	BPR	Vendor/Software Selection	IT Infrastructure	Compatibility	Relative Advantages	IT Expertise	Complexity	Security
		Dominant factors: Organizational factors						*IT-related factors*					*IT critical factors*												
32	Del Monte Foods		X		X			X	X	X						X				X	X		X	X	X
33	Dell	X	X		X			X	X	X	X				X	X	X	X		X	X	X	X		
34	Dexia Bank		X				X		X		X	X				X	X		X	X	X		X		
35	Dublin Airport Authority		X						X	X		X			X	X	X	X	X	X	X	X		X	
36	e-Bay	X		X	X		X		X	X		X	X				X		X	X			X		
37	Egg													X		X	X						X		
38	EMS Group	X	X		X		X		X	X			X	X	X	X	X		X	X		X	X	X	X
39	Environmental Protection Agency (EPA)				X			X	X	X	X				X	X			X	X	X			X	
40	FedEx	X															X							X	
41	Ford Motor	X															X		X	X					
42	Fortis	X									X						X								

continued on following page

237

Table 4. Continued

continued on following page

No.	Name of organization	Top Management Support & Commitment	Communication	Evaluation	Championship	Change Management	User Involvement	Financial Resources	Project Management	Business Case	Awareness	Training	Organization Structure	Organization Culture	Clear Vision & Objective	Implementation	Strategic Planning	BPR	Vendor/Software Selection	IT Infrastructure	Compatibility	Relative Advantages	IT Expertise	Complexity	Security
43	Freightliner LLC								X						X	X	X			X	X		X		
44	Garanti Bank	X	X	X										X			X		X						
45	Global Financial Services Leader	X	X					X	X	X		X				X	X	X	X	X	X	X	X	X	X
46	Honeywell		X		X				X	X	X	X			X	X					X	X	X	X	X
47	IBM	X	X		X					X							X			X	X	X	X		
48	Intel		X		X			X		X									X						
49	ITT Industries				X			X			X		X		X	X	X		X	X			X		
50	Jardine Air Terminal Services (JATS)		X		X		X						X		X	X	X				X			X	
51	JSC RusHydro	X	X	X	X		X		X	X					X	X	X	X			X		X		
52	KeyCorp	X														X	X								
53	Kings College														X				X	X	X				X

238

Table 4. Continued

No.	Name of organization	Top Management Support & Commitment	Communication	Evaluation	Championship	Change Management	User Involvement	Financial Resources	Project Management	Business Case	Awareness	Training	Organization Structure	Organization Culture	Clear Vision & Objective	Implementation	Strategic Planning	BPR	Vendor/Software Selection	IT Infrastructure	Compatibility	Relative Advantages	IT Expertise	Complexity	Security
54	Komatsu Europe International (KEISA)	X	X		X	X	X		X	X	X	X			X	X		X	X	X	X	X		X	X
55	Korean Association of Bookstores	X			X			X	X	X					X	X						X		X	
56	Linksys	X			X									X					X	X		X			
57	Lloyds TSB	X			X										X	X	X	X	X				X	X	
58	Meridian Bancorp	X			X					X							X	X							
59	MERSCORP, Inc.	X	X	X					X	X	X	X			X	X	X	X	X	X	X	X	X		X
60	Metro Inc.	X	X						X	X	X	X			X	X	X	X	X	X	X	X			
61	Ministry of Economics & Labour	X	X					X	X		X				X	X				X	X	X	X	X	X
62	Ministry of Transportation	X	X							X						X			X	X	X	X			
63	Mitsui Oil and Gas	X	X		X			X	X	X			X		X		X	X		X	X		X		

IT critical factors — *Dominant factors / Organizational factors / IT-related factors*

continued on following page

239

Table 4. Continued

No.	Name of organization	Top Management Support & Commitment	Communication	Evaluation	Championship	Change Management	User Involvement	Financial Resources	Project Management	Business Case	Awareness	Training	Organization Structure	Organization Culture	Clear Vision & Objective	Implementation	Strategic Planning	BPR	Vendor/Software Selection	IT Infrastructure	Compatibility	Relative Advantages	IT Expertise	Complexity	Security
64	Monsanto	X			X			X				X				X	X		X	X				X	
65	Mövenpick	X						X	X	X						X			X	X				X	X
66	NAV	X	X	X				X	X	X						X			X	X	X			X	X
67	NCCC	X														X	X	X	X	X	X	X	X		
68	NedTrain	X	X	X	X	X	X			X		X			X	X		X		X					
69	OCBC Bank	X	X		X				X	X	X	X			X	X	X		X	X	X	X	X		X
70	Panasonic Europe		X												X	X		X	X			X			
71	Pennsylvania Office of the Attorney General	X	X	X	X			X	X	X					X	X	X	X	X	X	X	X		X	
72	Philips Semiconductor	X		X						X						X	X	X	X	X					
73	Pondicherry	X	X	X				X				X			X					X		X	X	X	
74	Province of Genoa	X	X		X		X	X	X		X				X						X	X	X	X	X

continued on following page

240

Table 4. Continued

No.	Name of organization	Dominant factors — Organizational factors						IT-related factors					IT critical factors												
		Top Management Support & Commitment	Communication	Evaluation	Championship	Change Management	User Involvement	Financial Resources	Project Management	Business Case	Awareness	Training	Organization Structure	Organization Culture	Clear Vision & Objective	Implementation	Strategic Planning	BPR	Vendor/Software Selection	IT Infrastructure	Compatibility	Relative Advantages	IT Expertise	Complexity	Security
75	RailAmerica	X	X		X		X			X					X	X	X		X	X		X		X	X
76	SARA									X									X		X		X	X	
77	SaskEnergy	X	X					X	X						X	X	X		X	X	X	X			
78	Sharp Electronics Corporation	X	X		X			X	X	X	X				X	X	X		X		X	X	X	X	
79	Silicon Operations	X	X		X			X	X	X		X				X	X			X	X	X	X	X	
80	Statistics New Zealand	X	X		X		X	X	X	X	X				X	X	X		X	X	X	X		X	X
81	Syracuse Police Department	X	X		X	X		X						X			X	X		X	X		X		
82	TASER	X	X		X		X	X	X	X		X				X	X				X	X	X	X	X
83	Technogym		X	X	X											X			X	X	X	X		X	X
84	The City of Amers-foort	X			X										X				X	X	X		X	X	

continued on following page

241

Table 4. Continued

No.	Name of organization	Dominant factors / Organizational factors / IT-related factors — IT critical factors																							
		Top Management Support & Commitment	Communication	Evaluation	Championship	Change Management	User Involvement	Financial Resources	Project Management	Business Case	Awareness	Training	Organization Structure	Organization Culture	Clear Vision & Objective	Implementation	Strategic Planning	BPR	Vendor/Software Selection	IT Infrastructure	Compatibility	Relative Advantages	IT Expertise	Complexity	Security
85	The Highland Council	X	X	X	X			X		X		X			X	X	X		X	X	X	X	X	X	X
86	The Royal Air Force	X	X				X		X		X				X	X	X			X	X		X		X
87	Transtec			X		X											X		X	X					
88	Trimm Technologies	X	X					X								X		X	X	X		X		X	
89	TriView	X	X	X	X	X		X	X		X	X	X	X			X		X	X					
90	University of Michigan				X				X	X															
91	University of Texas	X			X					X										X					
92	US Agency	X	X	X	X	X	X	X	X				X	X	X	X	X	X	X	X	X	X	X	X	X
93	Veneto Region	X	X		X			X	X		X	X					X				X	X	X	X	X
94	Welch Allyn	X	X		X			X	X	X					X	X	X	X	X	X		X	X		

continued on following page

242

Table 4. Continued

No.	Name of organization	Top Management Support & Commitment	Communication	Evaluation	Championship	Change Management	User Involvement	Financial Resources	Project Management	Business Case	Awareness	Training	Organization Structure	Organization Culture	Clear Vision & Objective	Implementation	Strategic Planning	BPR	Vendor/Software Selection	IT Infrastructure	Compatibility	Relative Advantages	IT Expertise	Complexity	Security
95	West Marine	X	X												X				X						X
96	Westminster City Council	X	X		X												X		X		X	X	X	X	X
97	Whirlpool	X						X					X		X					X	X	X		X	
98	William Beaumont Hospital			X	X				X	X						X			X	X					
99	Wilton	X	X						X	X		X			X	X	X	X	X	X	X	X	X		
100	Woolwich	X	X						X						X		X		X			X	X	X	

(IT critical factors — Dominant factors / Organizational factors / IT-related factors)

Figure 4. Degree of criticality of critical factors in IT adoption

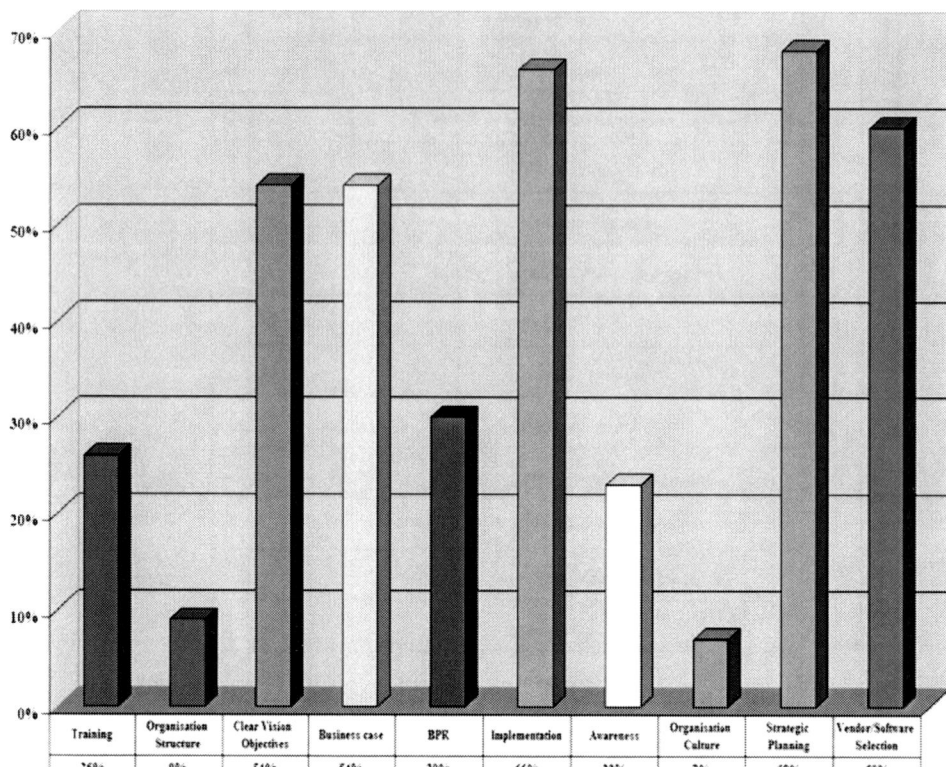

Figure 5. Degree of criticality of IT dominant factors

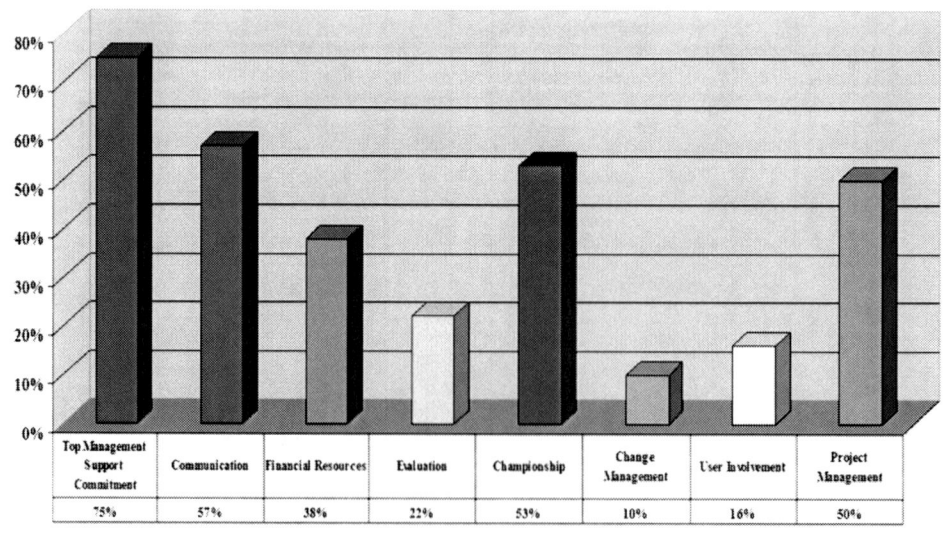

Finally, user involvement and change management were found to be less significant factors, with 16% and 10% of case studies highlighting them. One surprising result is that change management comes at the end. However, the literature reviewed emphasized that change management is one of the crucial factors in IT adoption projects (Poon and Wagner, 2001; Motwani, Subramanian and Gopalakrishna, 2005; Motawa et al., 2007). Change management is essential for preparing an organization for the introduction of an IT system and for its successful adoption.

Organizational Factors

Under this dimension, Figure 6 illustrates that 68% of case studies regard strategic planning as a critical factor in their IT adoption. Many researchers agree that clarity of IT strategy plan is positively associated with IT adoption success (Paulson, 1995; Teo et al., 1997; Tucker and Mohamed, 1996; Mak, 2001; Hwang, Ku, Yen and Cheng, 2004). The second significant organizational factor is implementation, with 66%. According to Vadapalli and Mone (2000): "The ability of an organization to execute and deliver on IT represents the implementation stage in the life cycle of

Figure 6. Degree of criticality of IT organizational factors

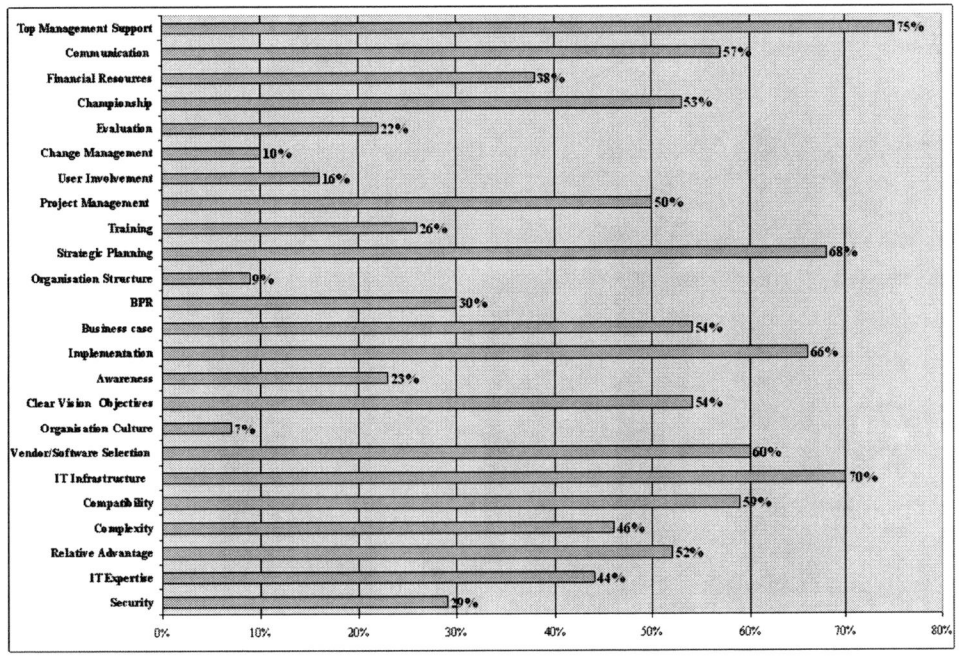

IT, and success in this stage leads to the actualization of benefits from IT". Next in terms of high criticality is vendor/software selection with 60%. Selection of vendors is generally a vital issue in IT projects (Shah and Siddiqui, 2006).

In addition, Figure 6 shows that two factors: clear vision and objectives and business case are the third most critical factors, with 54% each, while BPR, training and awareness come next with 30%, 26% and 23%, respectively. Finally, organization structure and culture were found to be less significant factors, evident in 9% and 7% of case studies.

IT-Related Factors

Figure 7 shows that the most critical factor in IT adoption in this dimension, which stands out over all others, is IT infrastructure, cited in 70% of case studies. This factor comes second after top management support and commitment. All organizations need a basic level of IT infrastructure to successfully adopt a new technology (Cash et al., 1992; Broadbent et al., 1999). This is confirmed by many researchers (Bharadwaj, 2000; Kayworth and Sambamurthy, 2000; Ragu-Nathan et al., 2004; Chanopas et al., 2006; Kim and Lee, 2007).

In addition, Figure 7 shows that compatibility is the second most significant critical factor in IT adoption in this dimension, with 59%. Also, 52% of these case studies considered relative advantage to be a critical factor in IT adoption, whereas

Figure 7. Degree of criticality of IT-related factors

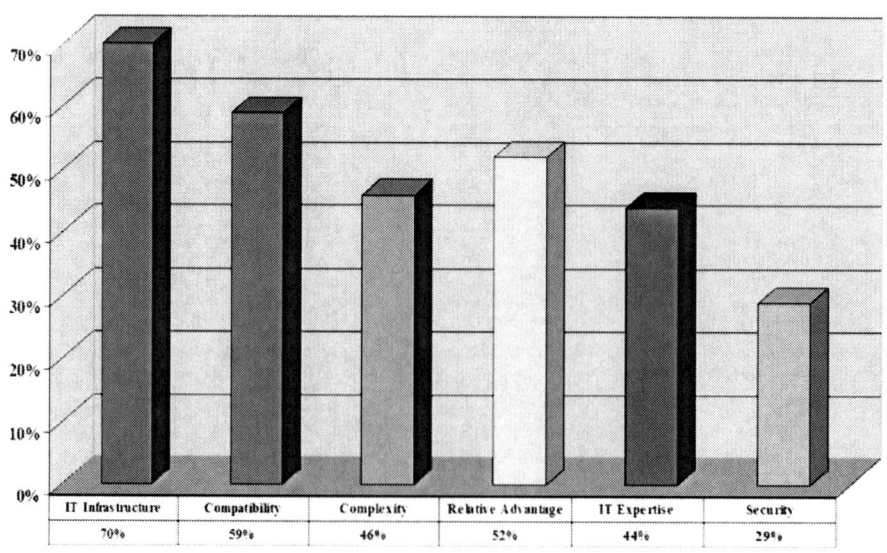

IT Infrastructure	Compatibility	Complexity	Relative Advantage	IT Expertise	Security
70%	59%	46%	52%	44%	29%

complexity and IT expertise came next occurring in 46% and 44% of organizations. On the other hand, security was seen as a critical factor in IT adoption by 29% of organizations.

SUMMARY AND FINDINGS

To attain the desired benefits, organizations must take into account several critical factors which contribute to the success of its adoption (Brandon, 2006). To do so, organizations need to be aware of the critical factors affecting the adoption process of an IT project. Brandon (2006) argues that "Completing IT projects successfully the first time requires the identification and understanding of all the critical success factors". Based on these arguments, this study has provided a detailed analysis of the critical factors essential for IT adoption in an organization. Twenty-four critical factors have been proposed based on the 100 different case studies. These factors correspond to the model for IT adoption in Figure 1.

Table 5. Degree of criticality of critical factors in IT adoption

No.	Factor	Percentage
1	Top management support & commitment	75%
2	IT infrastructure	70%
3	Strategic planning	68%
4	Implementation	66%
5	Vendor/software selection	60%
6	Compatibility	59%
7	Communication	57%
8	Clear vision & objectives	54%
9	Business case	54%
10	Championship	53%
11	Relative advantage	52%
12	Project management	50%
13	Complexity	46%
14	IT expertise	44%
15	Financial resources	38%
16	BPR	30%

continued on following page

Table 5. Continued

No.	Factor	Percentage
17	Security	29%
18	Training	26%
19	Awareness	23%
20	Evaluation	22%
21	User involvement	16%
22	Change management	10%
23	Organization structure	9%
24	Organization culture	7%

This chapter reveals that the critical factors are important and applicable. Organizations investigated gave considerable attention to most of the critical factors associated with IT implementation. The factor that stands out above all others as the most critical factor in IT adoption is the support and commitment of top management, closely followed by the IT infrastructure.

Table 5 shows the degree of criticality of critical factors in IT adoption. The ranking of these factors stemmed from the analysis of these case studies and the number of times their authors mentioned the factors.

REFERENCES

AIT. (2001). *Case study of Key Corp (NYSE: KEY)*.

AIT. (2001). *Case study of the Lloyds TSB Group*.

AIT. (2001). *Case study of the Woolwich*.

Al-Mashari, M. (2002). Electronic commerce: A comparative study of organizational experiences. *Benchmarking: An International Journal, 9*(2), 182–189. doi:10.1108/14635770210421845

Amazon. (n.d.). *Website*. Retrieved from www.Amazon.com

Avgerou, C., & Cornford, T. (1998). *Developing Information Systems: Concepts, issues and practice*. Palgrave Publisher.

Awallmah,N. A. H.(2002) Electronic government and the future of management: an empirical study in the public sector in Qatar country, magazine studies.

Baets, W. (1996). Some empirical evidence on IS strategy alignment in banking. *Information & Management, 30*(4), 155–177. doi:10.1016/0378-7206(95)00056-9

Baldrige National Quality Program. (2001). *TriView national bank case study*. Case study of TriView.

Bharadwaj, S. (2000). A resource-based perspective on information technology capability and firm performance: an empirical investigation. *Management Information Systems Quarterly, 24*(1), 169–197. doi:10.2307/3250983

Booty, F. (1998). Network management: The bottom line. *Manufacturing Computer Solutions, 4*(5), 37–40.

Bowonder, B., Akshay, J., & Narendra, K. (2000). *E-governance in a fisherman's community: A case study of Pondicherry*. Case Study of Pondicherry.

Brandon, D. (2006). *Project management for modern Information Systems*. USA: IRM Press.

Broadbent, M., Weill, P., & St Clair, D. (1999). The implications of Information Technology infrastructure for business process redesign. *Management Information Systems Quarterly, 23*(2), 159–182. doi:10.2307/249750

Bruque, S., & Moyano, J. (2007). Organizational determinants of information technology adoption and implementation in SMEs: The case of family and cooperative firms. *Technovation, 27*(5), 241–253. doi:10.1016/j.technovation.2006.12.003

Caroline, C., & Paul, M. (2000). *From EDI to Internet commerce; the BHP Steel experience*. Retrieved from www.emerald-library.com

FNS Case. (2001). *How a new approach to IT turned around the fortunes of Taiwan's Chinatrust Commercial Bank*. Case study of Chinatrust Commercial Bank.

Cash, J. I., McFarlan, W. F., McKenney, J. L., & Applegate, L. M. (1992). *Corporate Information Systems management: Text and cases* (3rd ed.). Homewood, IL: Irwin.

Chan, S. (1997). In search of Democratic peace: Problems and promise. *Mershon International Studies Review, 41*, 59–91. doi:10.2307/222803

Chanopas, A., Krairit, D., & Khang, D. B. (2006). Managing information technology infrastructure: A new flexibility framework. *Management Research News, 29*(10), 632–651. Emerald Group Publishing Limited. doi:10.1108/01409170610712335

Checkland, P., & Holwell, S. (1998). *Information, systems and information systems*. Chichester, UK: John Wiley.

Cisco. (2001). *Cisco and Blue Nile*. Retrieved from http:// www.cisco.com/warp/ public1779/ibs/solutions/ ecommercelbluenile cp.pdf

Clarke, S., & Doherty, N. (2004). The importance of a strong business-IT relationship for the realisation of benefits in e-business projects: The experiences of Egg. *Qualitative Market Research; Bradford, 7*(1), 58-66.

Claudia, L., & Stefan, S. (2001). *Web portfolio based electronic commerce: The case of Transtec AG*. Retrieved from www.emerald-library.com/ft

Clegg, C., Axtell, C., Damodaran, L., Farbey, B., Hull, R., & Lloyd-Jones, R. (1997). Information Technology: A study of performance and the role of human and organizational factors. *Ergonomics, 40*(9), 851–871. doi:10.1080/001401397187694

Damanpour, F. (1991). Organizational innovation: A meta-analysis of effects of determinants and moderators. *Academy of Management Journal, 34*(3), 555–590. doi:10.2307/256406

David, S. (2001). *eBay creates technology architecture for the future*. Sun Microsystems.

Dehning, B., Richardson, V., & Stratopoulos, T. (2005). Information technology investments and firm value. *Information & Management, 42*(7), 989–1008. doi:10.1016/j.im.2004.11.003

Dell. (2005, 2006, 2007 & 2008). *Customer case studies*. Dell. Retrieved from http:// www.dell.com/content/topics/ global.aspx/casestudies/

Dewett, T., & Jones, G. R. (2001). The role of information technology in the organization: A review, model and assessment. *Journal of Management, 27*(3), 313–346.

EDS. (2007 & 2008). *Case studies*. EDS. Retrieved from http://www.eds.com/ services/casestudies/

Farabi case. (2002). *Farabi connects remote customers and branch offices to corporate host for Caterpillar equipment dealer*. Case study of Al-Bahar.

FedEx. (n.d.). *Home page*. Retrieved from www.FedEx.com

Fujitsu. (2000). *Case studies*. Fujitsu. Retrieved from http://www.fujitsu.com/ global/casestudies/

Gartner Group. (2001). *Home page*. Retrieved from http://www.gartner.com/technology/home.jsp

Grover, G. (1998) *Identification of factors affecting the implementation of data warehousing*. Unpublished PhD Dissertation. Auburn University at Montgomery.

Hwang, H. G., Ku, C. Y., Yen, D. C., & Cheng, C. C. (2004). Critical factors influencing the adoption of data warehouse technology: A study of the banking industry. *Decision Support Systems, 37*, 1–21. doi:10.1016/S0167-9236(02)00191-4

IBM. (2001, 2006, 2007 & 2008). *Case studies.* IBM. Retrieved from http://www-01.ibm.com/software/success/

Kayworth, T., & Sambamurthy, V. (2000). Managing the information technology infrastructure. *Baylor Business Review, 18*(1), 13–15.

Keil, M., Mann, J., & Rai, A. (2000). Why software projects escalate: An empirical analysis and test of four theoretical models. *Management Information Systems Quarterly, 24*(4), 631–664. doi:10.2307/3250950

Kim, B. G., & Lee, S. (2007). Factors affecting the implementation of electronic data interchange in Korea. *Computers in Human Behavior, 24*(2), 263–283. doi:10.1016/j.chb.2006.11.002

Kumar, R. (2002). Managing risks in IT projects: An options perspective. *Information & Management, 40*, 63–74. doi:10.1016/S0378-7206(01)00133-1

Kwon, T. H., & Zmud, R. W. (1987). Unifying the fragmented models of information systems implementation. In Borland, R. J., & Hirschhiem, R. A. (Eds.), *Critical issues in information systems research* (pp. 252–257). New York, NY: John Wiley.

Lauria, E., & Duchessi, P. (2007). A methodology for developing Bayesian networks: An application to information technology (IT) implementation. *European Journal of Operational Research, 179*(1). Elsevier Science.

Lyytinen, K., & Hirschheim, R. (1987). Information failures—A survey and classification of the empirical literature. *Oxford Surveys in Information Technology, 4*, 257–309.

Mak, S. (2001). A model of information management for construction using information technology. *Automation in Construction, 10*, 257–263. doi:10.1016/S0926-5805(99)00035-7

McFadden, F. R. (1996). Data warehouse for EIS: Some issues and impacts. *Proc. of the 29th Hawaii International Conference on System Sciences* (pp. 120-129).

Microsoft. (2007 & 2008). *Case studies.* Retrieved from http://www.microsoft.com/casestudies/

Milis, K., & Mercken, R. (2004). The use of the balanced scorecard for the evaluation of Information and Communication Technology projects. *International Journal of Project Management, 22*, 87–97. doi:10.1016/S0263-7863(03)00060-7

Motawa, I. A., Anumba, C. J., Lee, S., & Pena-Mora, F. (2007). An integrated system for change management in construction. *Automation in Construction, 16*, 368–377. doi:10.1016/j.autcon.2006.07.005

Motwani, J., Subramanian, R., & Gopalakrishna, P. (2005). Critical factors for successful ERP implementation: Exploratory findings from four case studies. *Computers in Industry, 56*, 529–544. doi:10.1016/j.compind.2005.02.005

O'Brien, M. J., & Al-Biqami, N. M. (1999). Information Technology and the structure of the Saudi Arabian construction industry. *8th International Conference on Durability of Building Materials and Components*, (vol. 4, pp. 2327-2337). Ottawa, Canada: NRC Research Press, National Research Council of Canada.

Osei-Brysona, K. M., & Kob, M. (2004). Exploring the relationship between information technology investments and firm performance using regression splines analysis. *Information & Management, 42*, 1–13. doi:10.1016/j.im.2003.09.002

Paulson, B. (1995). *Computer applications in construction*. Singapore: McGraw-Hill Inc.

Poon, P., & Wagner, C. (2001). Critical success factors revisited: Success and failure cases of Information Systems for senior executives. *Decision Support Systems, 30*(3), 393–418. doi:10.1016/S0167-9236(00)00069-5

Premkumar, G., & Ramamurthy, K. (1995). The role of inter-organizational and organizational factors in the decision mode for adoption of inter-organizational systems. *Decision Sciences, 26*(3), 303–336. doi:10.1111/j.1540-5915.1995.tb01431.x

Premkumar, G., & Roberts, M. (1999). Adoption of new information technologies in rural small businesses. *Omega International Journal of Management Science, 27*(4), 467–484. doi:10.1016/S0305-0483(98)00071-1

Ragu-Nathan, B. S., Apigian, C. H., Ragu-Nathan, T. S., & Tu, Q. (2004). A path analytic study of the effect of top management support for information systems performance. *Omega, 32*, 459–471. doi:10.1016/j.omega.2004.03.001

Rajiv, D., & Robert, J. (1998). Case study of electronic banking at Meridian Bancorp. *Information and Software Technology, 33*(1), 1–9.

Ranganathan, C., & Kannabiran, G. (2004). Effective management of information systems function: An exploratory study of Indian organizations. *International Journal of Information Management, 24*, 247–266. doi:10.1016/j.ijinfomgt.2004.02.005

Ratnasingam, P. (2001). Inter-organizational trust in EDI adoption: The case of Ford Motor Company and PBR Limited in Australia. *Internet Research; Bradford, 11*(3), 261-268.

Rodriguez-Repiso, L., Setchi, R., & Salmeron, J. (2007). Modelling IT projects success with Fuzzy Cognitive Maps. *Expert Systems with Applications*, *32*, 543–559. doi:10.1016/j.eswa.2006.01.032

Rogers, E. (1995). *Diffusion of innovations* (4th ed.). New York, NY: Free Press.

Saleh, Y., & Alshawi, M. (2005). An alternative model for measuring the success of IS projects: The GPIS model. *The Journal of Enterprise Information Management*, *18*(1), 47–63. doi:10.1108/17410390510571484

Sauer, C., & Cuthbertson, C. (2003). *The state of IT project management in the UK.* Oxford: Templeton College.

Shah, M. H., & Siddiqui, F. A. (2006). Organizational critical success factors in adoption of e-banking at the Woolwich bank. *International Journal of Information Management*, *26*, 442–456. doi:10.1016/j.ijinfomgt.2006.08.003

Slevin, D. P., & Pinto, J. K. (1986). The project implementation profile: New tool for project managers. *Project Management Journal*, *17*(4), 57–70.

Soliman, K. S., & Janz, B. D. (2004). An exploratory study to identify the critical factors affecting the decision to establish Internet-based interorganizational information systems. *Information & Management*, *41*, 697–706. doi:10.1016/j.im.2003.06.001

Swanson, E. B., & Wang, P. (2005). Knowing why and how to innovate with packaged business software. *Journal of Information Technology*, *20*, 20–31. doi:10.1057/palgrave.jit.2000033

Tarafdar, M., & Vaidya, S. (2006). Challenges in the adoption of e-commerce technologies in India: The role of organizational factors. *International Journal of Information Management*, *26*, 428–441. doi:10.1016/j.ijinfomgt.2006.08.001

Teo, T. S. H., Ang, J., & Pavri, F. (1997). The state of strategic IS planning practices in Singapore. *Information & Management*, *33*, 13–23. doi:10.1016/S0378-7206(97)00033-5

Thong, J. Y. L. (1999). An integrated model of information system adoption in small business. *Journal of Management Information Systems*, *15*(4), 187–214.

Tucker, S. N., & Mohamed, S. (1996). Introducing information technology in construction: Pains and gains. *Proceedings of the CIB-W65 Symposium on Organization and Management of Construction,* Glasgow, (pp. 348-356).

Turban, E., Lee, J., King, D., & Shung, H. M. (2000). *Electronic commerce, a managerial perspective.* London, UK: Prentice Hall.

Vadapalli, A., & Mone, M. A. (2000). Information technology project outcomes: User participation structures and the impact of organization behavior and human resource management issues. *Journal of Engineering and Technology Management, 17*, 127–151. doi:10.1016/S0923-4748(00)00018-7

Willcocks, L., & Lester, S. (1994). Information Systems investments: Evaluations at the feasibility stage of project. *Technovation, 11*(5), 283–302. doi:10.1016/0166-4972(91)90027-2

Wilson, R. (2006). *New digital state leaders emerge from e-government study.* Center for Digital Government. Retrieved from www.centerdigitalgov.com/ center/ highlighstory.html

Zee, H. V. D. (2002). *Measuring the value of Information Technology.* Hershey, PA: Idea Group Publishing.

Chapter 11
Intelligent Decision Making and Risk Analysis for IT Management Processes

Masoud Mohammadian
University of Canberra, Australia

Ric Jentzsch
University of Canberra, Australia

EXECUTIVE SUMMARY

IT management processes have been growing as the development of modern IT systems has grown. These are often complex with multiple interdependencies that can make it very difficult for Chief Information Officers (CIOs) to comprehend and be aware of potential risks. These risks have the potential to translate into decision making inefficiencies for an organization. Risk analysis for decision making in the planning and monitoring of these systems can be a complex and demanding task. Intelligent decision making in IT management processes and systems are a crucial element of an organization's success and its competitive position in the marketplace. This chapter considers the implementation of Fuzzy Cognitive Maps (FCM) to provide facilities to capture and represent complex relationships in an IT management process model. By using FCMs, CIOs can regularly review and improve their IT management processes and provide greater improvement in development, monitoring and maintenance of those processes. CIOs can perform what-if analysis to better understand vulnerabilities of their designed system.

DOI: 10.4018/978-1-61350-311-9.ch011

INTRODUCTION

Decision making can proceed by informal deliberation or with the help of an analytical technique of different types (Van Gelder, 2010). Deliberation is the careful consideration of options and related issues expressed as relevant arguments and evidence (Van Gelder, 2010). Everyone has and uses informal deliberation in some type of decision making process, which may be habitual, unconscious, or by default. The informal process is automatically adopted when no other decision making process has been selected (Van Gelder, 2010). With deliberation in decision making comes risk.

There is an urgent need to improve decision making in IT management. Informal decision making methods can be unreliable, and can lead to bad decisions (where the wrong choice was made), based on thinking that was clearly ill-informed, sloppy, disorganised, incomplete or biased (Van Gelder, 2010). Analytical techniques for decision making, though employing a disciplined and systematic approach, can partly be developed to overcome the unreliability of informal methods (Van Gelder, 2010). However even these are not exempt from the possibility of establishing bad decisions.

Decision making in the real world is frequently dynamic and often constrained by limited or restricted resources. Real world decision situations exist in the context of systems, which are usually characterised by a number of interacting concepts that evolve over time (Salmeron, 2009). Any support to helping the decision makers reduce the risk of incorrect decisions will provide long term benefits to an organization. To do this, decision makers need to construct a representation of the decision problem, establish alternative courses of action, and imagine or calculate the outcome of choosing an alternative (Salmeron, 2009). This chapter provides just such a model in analysing IT management processes and developing a risk analysis based on using FCM with the model.

INFORMATION TECHNOLOGY PROCESSES

IT processes are activities for development and maintenance of applications, supporting infrastructure (e.g., hardware, systems software, and networks), to managing human resources. Luftman, Bullen et al., (2004) described 38 IT processes that cover all aspects or IT management in an organization. These IT processes have been categorized in three main layers. These layers include:

1.   Strategic - focus is on long-term goals, for SME's that is approximately 3 to 4 years, while for large enterprises it is ≥ 4 years;

Figure 1. Strategic layer sub-processes

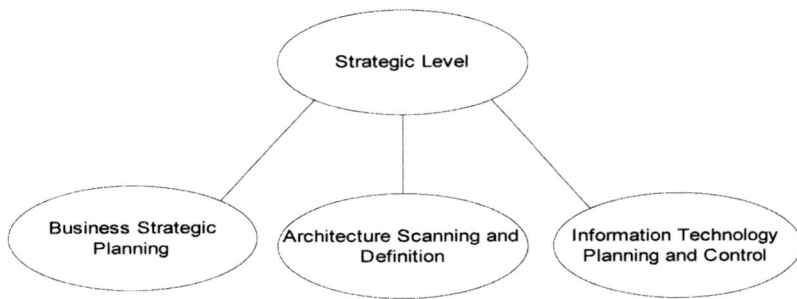

2. Tactical - focus is on yearly activities, for SMEs and large enterprises approximately one year. Focus is also on supporting and feeding into the strategic layer; and

3. Operational - for SMEs and large enterprises this is the daily to monthly activities. This layer supports and feeds into the tactical layer.

The strategic layer consists of strategic planning that covers business analysis planning, architecture planning and IT strategic planning control. Figure 1 shows sub-processes of the strategic layer.

The tactical layer consists of management planning, development planning, resource planning, financial planning, and service planning. These five processes can be further divided into 15 sub-processes. For this analysis we will concentrate on the five processes only. Figure 2 shows the sub-processes.

The operational level consists of six processes: project management, resource control, service control, development and maintenance, administration services and information services. These six processes could be divided into 22 sub processes. However, for this analysis we will concentrate on the six processes only. Figure 3 shows the processes of the operational layer.

Figure 2. Tactical layer sub-processes

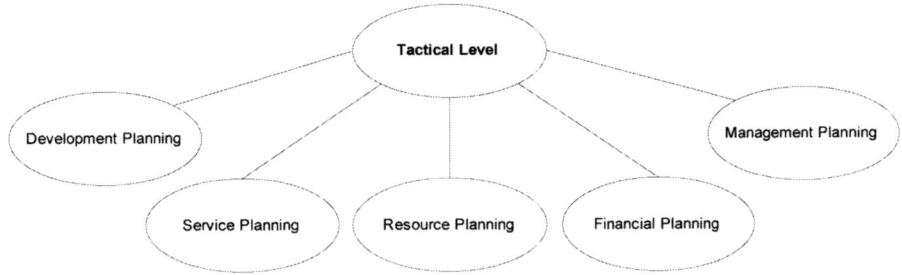

Figure 3. Operational layer sub-processes

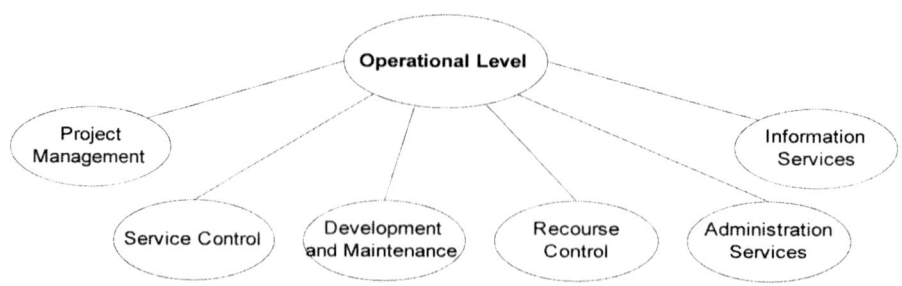

Using the three layer approach, it is possible to distinguish the sub-processes and codependences between IT functions. It can be noted that the strategic layer impacts the tactical layer by changing the technologies, tools, and methodologies used in tactical processes. Consequently new technologies, tools, and methodologies impact the operational level by changing the staff requirements, training needs, and job functions (Luftman, Bullen et al., 2004).

Research indicates that leading management-consulting firms such as Ernst & Young, Price Waterhouse Coopers, as well as the Society of Information Management (SIM) provide a different number of IT processes and sub-processes. For example Ernst & Young have presented 70 IT processes, Price Waterhouse Coopers use 62 IT processes, and the Society of Information Management (SIM) has listed 40 IT processes. CobiT and the IT Governance Institute (for example) use 34 IT processes. The CobiT model also consists of six major categories but there is no distinction between strategic, tactical, or operational layer.

No matter which model is used there are a large number of sub-processes that the Chief Information Office (CIO) needs to consider to be able to successfully manage the IT function for an organization. Other models can easily be substituted to evaluate the risk in IT management based on the particular model.

In the next section the three layers of the model identified by Luftman, Bullen et al., (2004) are further analyzed. Section 4 provides a brief overview of Fuzzy Cognitive Maps (FCMs) and section 5 provides simulation details of application of FCMs to this three layer IT management process. It provides the facilities that a CIO requires to perform risk analysis and simulate what-if scenarios that need to be considered during the development, implementation and monitoring of IT management processes using this three level model.

STRATEGIC, TACTICAL AND OPERATIONAL LAYER

The strategic layer consists of processes that have long term results on an organization. These processes provide a competitive edge for an organization by delivering cost efficiencies and improvements to business processes.

In contrast to the strategic layer, the tactical layer processes are relatively short-term. Their impact on an organization can be observed within a year, as improvements in existing operations. The tactical layer processes involve, in one way or another, the majority of the IT staff and budget.

Operational layer processes cover day-to-day, month-to-month operations and functions that are critical to an organization. CIOs and IT executives manage and monitor IT systems and conduct IT management to improve an organization's IT efficiencies and effectiveness. Some of the questions that they need answers to on a regular basis are:

Which IT processes are most important? How does any change in these processes affect the importance of the remaining processes?

Who owns each of these processes? How many resources will need to be applied to each process? How efficient are each of these processes today? What priority should be placed on improving each of these processes?

How will the IT strategic processes have an impact on the pace of technological change in an enterprise?

What is the role of the IT architecture in aligning businesses and IT functions of an organization?

Effective IT management depends on the success and application of IT processes. Some processes have more important roles and impact than other IT processes. Development of effective applications in an organization will provide the basis for the organization to be placed into a more advantaged competitive and strategic position.

IT departments and the CIO need to take ownership of IT processes and develop these processes efficiently and effectively. Resource allocation for each IT process should be based on the importance and complexity of such processes. For example operational level processes provide more than 60 percent (Luftman, Bullen et al., 2004) of IT activities in an organization. However, less than 40 percent of the IT staff are allocated on operational activities. With careful consideration and use of new methodologies many IT processes can be improved. Improving the IT process is costly in terms of budget and time. IT executives are trying to reduce costs while there is a need to increase IT budgets that will subsequently reduce or provide cost advantages to the organization.

Finally effective management and performance of IT can provide a successful edge for an organization. The ranking and prioritization of IT processes are difficult tasks and require leadership and effort from CIOs and IT executives.

The next section provides a brief overview of Fuzzy Cognitive Maps (FCMs) and proposes FCM for performing risk analysis and simulating what-if scenarios that can be conducted during the development of IT management processes using this three level model.

FUZZY COGNITIVE MAPS

Fuzzy Cognitive Maps (FCM) (Luftman, Bullen et al., 2004) are graph structures that provide a method of capturing and representing complex relationships in a system. The application of FCM has been popular in modeling problems with low or no past data set or historical information (Kosko, 1997; Kosko, 1986; Aguilar, 2005; Georgopoulous, Malandrak et al., 2002; Andreou, Mateou et al., 2003; Axelrod, R. 1976; Carlsson & Fuller, 1996; Tsadiras, Kouskouveliset al., 2001;). A FCM provides the facilities to capture and represent complex relationships in a system to improve the understanding of a system. A FCM uses scenario analysis by considering several alternative solutions to a given situation. Concepts sometimes called nodes, elements, or events represent the system behavior in a FCM. These concepts are connected using a directed arrow showing causal relations between concepts. The graph's edges are the casual influences between the concepts. The development of the FCM is based on the utilization of domain experts' knowledge. Expert knowledge is used to identify concepts and the degree of influence between them.

Kosko enhanced cognitive maps by including fuzzy values for the relationships between concepts (Kosko, 1997, and Kosko, 1986). FCM applications have been very popular in modeling for problems with low or no past data set or historical information. FCM allows capturing and representing complex relationships (Kosko, 1997; Kosko, 1986; Aguilar, 2005; Georgopoulous, Malandrak et al., 2002; Andreou, Mateou et al., 2003; Axelrod, 1976; Carlsson, Fuller 1996; Tsadiras, Kouskouvelis et al., 2001). A FCM describes a system as a directed graph. The concepts are connected using a directed arrow showing causal relations between concepts. The graph's edges are the casual influences between the concepts. The value of a node reflects the degree to which the concept is active in the system at a particular time.

This value is a function of the sum of all incoming edges multiplied and the value of the originating concept at the immediately preceding state. A threshold function applied to the weighted sums. Values on each edge indicate relationships between concepts. These values indicate whether one concept increases or decreases the likelihood of another concept. The edges have values in the interval range [–1, 1]. These values indicate the degree to which one concept affects another. A positive relationship between concept 1 and concept 2 indicates an increase in the likelihood of concept 2 occurring. Negative values indicate a decrease in the likelihood of concept 2 occurring. The FCM represents the sub-processes in each layer of the IT management and monitoring model.

A FCM can be used in conjunction with possibilities for what-if analysis by CIOs to understand the vulnerability in each of the layers of their model. FCM can also be used to do risk analysis thereby providing the decision makers with additional information on which to base their IT management decisions. Using FCM it is possible to do what-if and scenario analysis to access the model. In this chapter a FCM is utilized to perform risk and scenario analysis in understanding vulnerabilities of the IT management model.

SIMULATION USING FCM FOR IT MANAGEMENT AND DECISION MAKING

The application of FCM to the IT Management processes begins by reviewing the relationships between the sub-processes. These relationships indicate whether one event/sub-process increases or decreases the likelihood of another event/sub process occurring (Kosko, 1997; Kosko, 1986; Aguilar, 2005; Georgopoulous, Malandrak et al., 2002; Andreou, Mateou et al., 2003; Axelrod, 1976; Carlsson & Fuller 1996; Tsadiras, Kouskouvelis et al., 2001). The sub-processes of strategic, tactical and operational layers can be converted into a FCM by considering sub-process to represent concepts of a FCM.

Figure 4 shows three FCM with weights allocated to each edge based on opinion of experts (Kosko, 1988; Taber, 1991).

CIOs and IT executives are the experts and are required to determine the weights on the different causal links and the initial activation level for each concept. In this scenario the author has carefully considered the system and provided the weights for the FCM as shown in Figure 4. The weights and the activation function value will vary for different organizations based on their budget, priorities and importance of sub-processes for the given organization.

The mathematical model behind the graphical representation of the FCM consists of a $1 x n$ state vector I. This state vector represents the values of the n concepts and $n x n$ weight matrix W_{IJ} represents value of weights between concepts of C_i and C_j. For each concept in a FCM a value one or zero is assigned. One represents the

Figure 4. FCM representation of sub-process for the strategic layer (a), tactical layer (b), and operational layer (c)

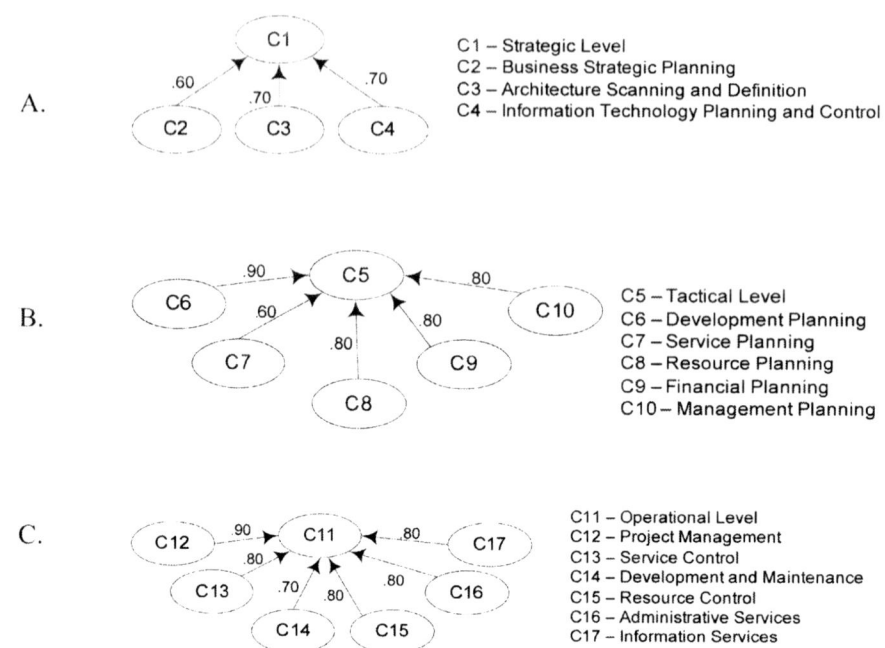

existence of that concept at a given time and zero represents non-existence of the respective concept. A threshold function is used in FCM. The threshold function used in this research study is sigmoid function (Kosko, *1997;* Andreou, Mateou et al., 2003; Axelrod, 1976; Carlsson & Fuller 1996; Tsadiras, Kouskouvelis et al., 2001).

$$C_i(t_{n+1}) = S\left[\sum_{K=1}^{N} e_{KI}(t_n)C_k(t_n)\right] \tag{1}$$

For example the relationships in Figure 4 between C5 (Tactical Level) and C9 (Financial Planning) implies that if financial planning increases, then it will influence the success of the tactical level, thus resulting in risk reduction. A CIO can evaluate, modify the concepts and weights of an FCM based on their resources and priority of IT processes in their organization.

FCM can be used to identify and access risks that may arise due to unavailability or shortcomings of different IT processes for a given organization. It should be emphasized the weights in FCM are calculated based on expert knowledge (CIOs,

IT executives and other decision makers) which shows the influences of concepts to other concepts. These weights can be changed based on CIOs and other expert's expertise, knowledge of the organization and availability of resources. The threshold function used in this research study is sigmoid function and its value is set to 0.2.

The relationship details among all concepts in Figure 4 for strategic, tactical and operational layers can be displayed using in a matrix form as follows.

$$E_{Strategic} = \begin{bmatrix} 0 & 0 & 0 & 0 \\ 0.6 & 0 & 0 & 0 \\ 0.7 & 0 & 0 & 0 \\ 0.7 & 0 & 0 & 0 \end{bmatrix}$$

$$E_{Tactical} = \begin{bmatrix} 0 & 0 & 0 & 0 & 0 \\ 0.9 & 0 & 0 & 0 & 0 \\ 0.6 & 0 & 0 & 0 & 0 \\ 0.8 & 0 & 0 & 0 & 0 \\ 0.8 & 0 & 0 & 0 & 0 \\ 0.8 & 0 & 0 & 0 & 0 \end{bmatrix}$$

$$E_{Operational} = \begin{bmatrix} 0 & 0 & 0 & 0 & 0 & 0 & 0 \\ 0.9 & 0 & 0 & 0 & 0 & 0 & 0 \\ 0.8 & 0 & 0 & 0 & 0 & 0 & 0 \\ 0.7 & 0 & 0 & 0 & 0 & 0 & 0 \\ 0.8 & 0 & 0 & 0 & 0 & 0 & 0 \\ 0.8 & 0 & 0 & 0 & 0 & 0 & 0 \\ 0.8 & 0 & 0 & 0 & 0 & 0 & 0 \end{bmatrix}$$

A what-if analysis can proceed by using the above matrices. For this simulation the threshold is set to be 0.2. Consider the following scenario:

- What happens if the event C9 (i.e. Financial Planning) is performed successfully?

This scenario can be presented using vector I_0 representing this situation by: I_0 = [0, 0, 0, 0, 1, 0]. In vector I_0 the concept C9 is represented as the fifth element in the vector and it is set to 1 and all other elements are set to zero representing that the other events have yet to take place. It is assumed that C9 happens and no other event has happened. Now I_o*E can provide the solution for this situation as follows: $I_I*E = [0.8, 0, 0, 0, 0, 0] = I_I$ which concludes that if C9 happens then it will increase the possibility of success in C5 (i.e. the tactical level to occur by 80%).

- Are the right resources being used and are they integrated properly in the tactical layer?

This scenario can be presented using vector I_o representing this situation by: $I_o = [0, 0, 0, 1, 0, 0]$. In vector I_o the concept C8 is represented as the fourth element in the vector and it is set to 1 and all other elements are set to be zero representing other events that have not happened. It is assumed that C8 happens and no other event has happened. Now I_o*E can provide the solution for this situation as follows: $I_I*E = [0.7, 0, 0, 0, 0] = I_I$ which concludes that if C8 happens then it will increase the possibility of success in C4 i.e. the tactical level to occur by 70%. Many other scenarios involving questions that can invoke several concepts at a given time can also be considered. Consider the following question:

- What happens if the event C9 (i.e. Financial Planning) is performed? Are the right resources being used?

This scenario can be presented using vector I_o representing this situation by: $I_o = [0, 0, 0, 1, 1, 0]$. In vector I_o the concept C8 and C9 are represented as the fourth and fifth elements in the vector and they are set to 1 and all other elements are set to be zero representing other events that have not happened. It is assumed that C8 and C9 happen and no other event has happened. Now I_o*E can provide the solution for this situation as follows: $I_I*E = [1.6, 0, 0, 0, 0, 0] = I_I$ which concludes that if C8 and C9 happen then it will increase the possibility of success in C5 (i.e. the tactical level).

Other 'what if' scenarios can easily now be performed on this FCM. Several simulations were performed using different scenarios. Table 1 displays the consequences of different scenarios based on different 'what if' simulations.

Using scenario analysis the CIOs can identify problems before they occur. Risk assessment and risk management requires proper analysis and understanding of the threats and shortcoming in an IT system. Currently there are no facilities that can be utilized to assess the above-mentioned three layer model and to mitigate risks at this level. FCM can help to fill this gap in risk assessment.

Table 1. The consequences of different scenarios based on different what-if simulations

'What if' the following scenario	Result of each scenario Consequences
C6	C6 $\xrightarrow{20\%}$ C5
C7	C7 $\xrightarrow{209\%}$ C5
C8	C4 $\xrightarrow{30\%}$ C5
C9	C9 $\xrightarrow{30\%}$ C5
C6 & C7	C6 & C7 $\xrightarrow{40\%}$ C5
C6 & C7 & C8	C6 & c7 & C8 $\xrightarrow{79\%}$ C5
C6 & C7 & C8 & C9	C6 & C7 & C8 & C9 $\xrightarrow{100\%}$ C5

SIMULATION OF COMBINED THREE LAYER ANALYSIS

The information provided from 'what-if' scenarios can be used for risk analysis by CIOs. To this point each layer has been considered separately. However, this approach can be used by CIOs to evaluate risks over the three layers. The CIOs can make decisions and manipulate resources for sub-processes, change tools available, affect the number of staff etc and re-evaluate risk for each change.

It is possible using FCM to produce an exhaustive list of analysis scenarios to evaluate risks. Risk analysis at different levels of IT management is useful for management at that level. However the interdependencies of this three level IT management process model provides an additional perspective on risk analysis based on all levels as an enterprise view. Figure 5 shows a FCM representation of the combined three levels of IT management processes.

'What-if' analysis can proceed using the matrix $E_{Strategic\,Tactical\,Operational} = E_{Strategic} + E_{Tactical} + E_{Operational}$.

The interdependencies C5 and C11 can be tested based on the FCM model shown in Figure 5. This situation can be presented using vector I_0 by:

$$I_0 = [0, 0, 0, 0, 1, 0, 0, 0, 0, 0, 1, 0, 0, 0, 0, 0, 0]$$

$$I_0 * E_{Strategic\,Tactical\,Operational} = [0.8, 0, 0, 0, 0.9, 0, 0, 0, 0, 0, 0, 0, 0, 0, 0, 0, 0] \text{ then } \rightarrow$$
$$I_1 = [1, 0, 0, 0, 1, 0, 0, 0, 0, 0, 1, 0, 0, 0, 0, 0, 0]$$

Figure 5. Combined three layers FCM for IT management

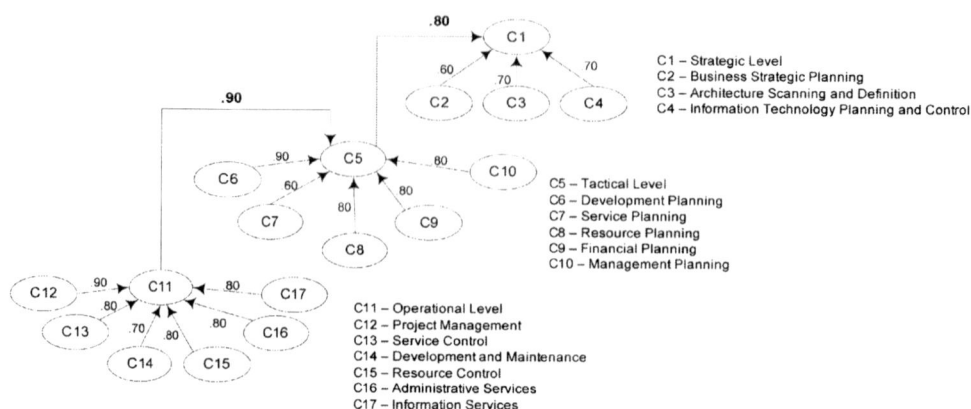

$I_1 * E$ *Strategic Tactical Operational* $= [0.8, 0, 0, 0, 0.9, 0, 0, 0, 0, 0, 0, 0, 0, 0, 0, 0, 0]$ then $\rightarrow$
I_2 $[1, 0, 0, 0, 1, 0, 0, 0, 0, 0, 1, 0, 0, 0, 0, 0, 0, 0, 0, 0]$

Thus if C5 and C11 happen, C1 will happen.

Regardless of the concepts and influence weights use, the analysis must precede along the flow of the arcs as indicated by the arrows. Table 2 shows several simulations (using the FCM model shown in Figure 5) where different concepts have been turned on while all others have remained off.

The information provided from the 'what-if' scenarios from all three layer IT management processes can be used for risk analysis by CIOs. To illustrate the risk analysis we will begin by using the event occurs row number 1 in Table 3.

Resource planning + service planning + development planning increase the probability of success at the tactical level. This then carries through to the strategic level increasing the possibility of success. Or to put it another way, there still exists a risk that even though C6, C7, and C8 meet their objectives, that the strategic level will still not be met.

Simulation of FCM for Time Delayed Decision Making for IT Management

In many applications the decision maker needs to not only know the effect of sub-processes to a system as described above but also they need to know the amount of time delay caused by malfunctioning of a sub-process in different layers of IT management (Luftman, Bullen et al., 2004). Decision making in the real world application is frequently dynamic and often constrained by limited or restricted

Table 2. FCM matrix for combined strategic, tactical, and operational layers

	C1	C2	C3	C4	C5	C6	C7	C8	C9	C10	C11	C12	C13	C14	C15	C16	C17
C1	0	0	0	0	0	0	0	0	0	0	0	0	0	0	0	0	0
C2	0.6	0	0	0	0	0	0	0	0	0	0	0	0	0	0	0	0
C3	0.7	0	0	0	0	0	0	0	0	0	0	0	0	0	0	0	0
C4	0.7	0	0	0	0	0	0	0	0	0	0	0	0	0	0	0	0
C5	0.8	0	0	0	0	0	0	0	0	0	0	0	0	0	0	0	0
C6	0	0	0	0	0.9	0	0	0	0	0	0	0	0	0	0	0	0
C7	0	0	0	0	0.6	0	0	0	0	0	0	0	0	0	0	0	0
C8	0	0	0	0	0.8	0	0	0	0	0	0	0	0	0	0	0	0
C9	0	0	0	0	0.8	0	0	0	0	0	0	0	0	0	0	0	0
C10	0	0	0	0	0.8	0	0	0	0	0	0	0	0	0	0	0	0
C11	0	0	0	0	0.9	0	0	0	0	0	0	0	0	0	0	0	0
C12	0	0	0	0	0	0	0	0	0	0	0.9	0	0	0	0	0	0
C13	0	0	0	0	0	0	0	0	0	0	0.8	0	0	0	0	0	0
C14	0	0	0	0	0	0	0	0	0	0	0.7	0	0	0	0	0	0
C15	0	0	0	0	0	0	0	0	0	0	0.8	0	0	0	0	0	0
C16	0	0	0	0	0	0	0	0	0	0	0.8	0	0	0	0	0	0
C17	0	0	0	0	0	0	0	0	0	0	0.8	0	0	0	0	0	0

Table 3. Consequences of different scenarios based different 'what if' simulations

	What if the event occurs	Consequences
1	C8, C7, C6	$C8, C7, C6 \xrightarrow{2.3} C5 \xrightarrow{.80} C1$
2	C15, C12, C11, C3	$C15, C12 \xrightarrow{1.70} C11 \xrightarrow{.90} C5, C3 \xrightarrow{1.50} C1$
3	C4, C7, C9, C10	$C10, C9, C7 \xrightarrow{2.20} C5, C4 \xrightarrow{1.50} C1$
4	C2, C8, C12, C14	$C14, C12 \xrightarrow{1.60} C11, C8 \xrightarrow{1.70} C5, C2 \xrightarrow{1.40} C1$
5	C7, C8, C9, C15	$C15 \xrightarrow{.80} C11, C7, C8, C9 \xrightarrow{3.10} C5 \xrightarrow{.80} C1$
6	C4, C10, C13, C15	$C15, C13 \xrightarrow{1.60} C11, C10 \xrightarrow{1.70} C5, C4 \xrightarrow{1.50} C1$
7	C7, C9, C12, C13	$C13, C12 \xrightarrow{1.70} C11, C9, C7 \xrightarrow{2.30} C5 \xrightarrow{.80} C1$

resources and time (Salmeron, 2009). Real world decision situations exist in the context of systems, which are usually characterised by a number of interrelated and interacting concepts that evolve over time (Salmeron, 2009). The decision makers are required to provide as much information as possible for them to be able to assist them in decision making. As such, time delays need to be considered in the above IT management processes as shown in Figure 5. This section considers fuzzy logic to represent time delays in IT management. Fuzzy logic can be applied to the above model in analysing IT management processes and developing a risk analysis based on using FCM as shown in Figure 5.

Understanding the time delay in IT management due to malfunctioning or unavailability of resources is significant information for an organization in determining and deploying proper decision making and risk assessment.

Such a time delayed model for IT management needs to be able to at least determine the estimated time delay value. When decision making is of paramount importance, time delay information should consider issues such as sub-processes involved in each layer of the IT management based on Luftman, Bullen et al., 2004.

An understanding of the nature of each process and its importance to the organization, its activities and usage of each sub-process in each layer can enhance organizational activities. It will also assist in allocating the appropriate time and resources to each sub-process. In such a situation, time allocation for each layer of the IT management can be classified. Based on such time classification, time intensive IT management layers can be determined and classified to "very high time consuming", "high time consuming", "moderate time consuming", "very low time consuming" and "low time consuming" layers.

Time classification of IT management layers is based on:

*   Understanding of tasks required to complete each layer as well as the required resources and times for each task,
*   Determining the importance of each layer and allocation of appropriate time and resources for each process.

Another issue related to time classification for each layer of IT management is defining the level of time classification for each layer. There is no exact and firm rule on the level of time classifications, however issues such as the size, type, level of IT resources used in an organization and corporate objective and regulatory rules are a few to consider.

Time classification can be expressed in human understandable language as they are mostly vague and difficult to estimate for IT layers and its associate processes. The excessive gap between precision of classic logic and imprecision and vagueness in definition of polices creates difficulty in representing these policies in formal logic.

Figure 6. FCM for IT management decision making with time delays as fuzzy rules information to each layer

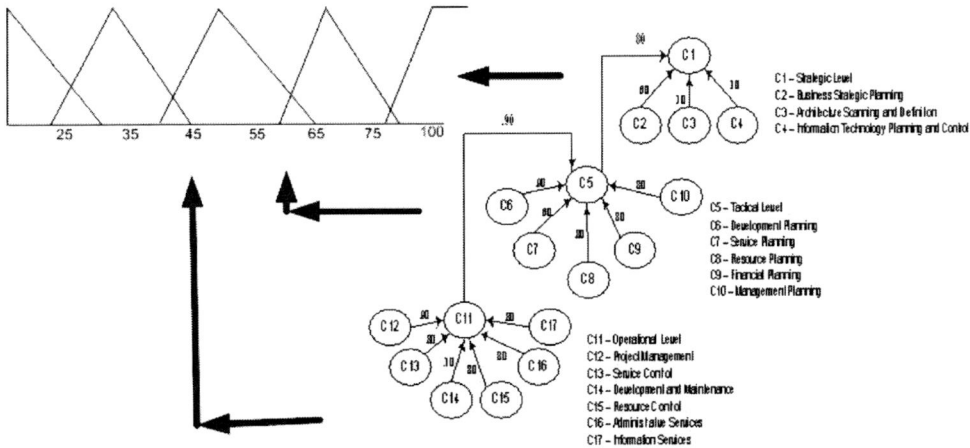

Zadeh's work (1965) has been found to be useful in its ability to handle vagueness. In this research study a time classification method is presented to determine time classification levels for IT management layers.

Time classification levels could divide time into classes such as "very high", "high", "moderate," "low" and "very low" time intensive layers. Based on these class categories, time usage for the IT management layers can be identified.

To perform such time classification with minimal resources impact and without needing to re-design FCM for IT management decision making, one option is to add extra information to each layer of IT management in FCM, shown in Figure 5, by adding time delays as fuzzy rules information to each IT management layer.

These fuzzy rules could be the value or degree of time delay for that IT management layer. This can be demonstrated as shown in Figure 6.

These fuzzy rules can then be used for adaptation and implementation of time classification for each IT management layer. The fuzzy rules can be created from the knowledge workers of the organization. Assume that the following domain values for these linguistic variables, VHT= very high level time consuming, HT= high level time consuming, MT = "medium level time consuming", LT ="low level time consuming", VLT = "very low level time consuming", The values related to linguistic variables are: VHT = [80,..,100], HT = [60,..,85], MT = [40,..,65], LT =[45,..,25], VLT = [0,..,30]. The time values are in the range of 0 to 100, where zero indicates a layer that is very low time consuming and 100 indicates a high time consuming layer. Note that other values can also be allocated.

Figure 7. Center of gravity inference method

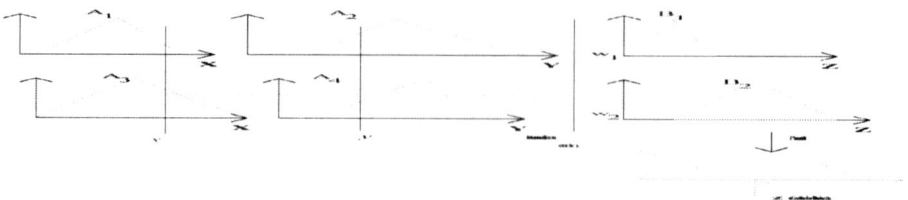

Based on the above values for each IT management layer, the membership of that IT management layer to each linguistic variable can be calculated. In Figure 7, triangular and trapezoidal fuzzy set was used to represent the time classifications (e.g. HT= high level time consuming, MT = "medium level time consuming", LT ="low level time consuming", VLT = "very low level time consuming"). The membership value of a layer can be calculated for all these time classifications using the following formulas:

$$m_A(x) = 0, x < a_1$$

$$m_A(x) = \frac{x - a_1}{a_2 - a_1}, a_1 \leq x \leq a_2$$

$$m_A(x) = \frac{a_3 - x}{a_3 - a_2}, a_2 \leq x \leq a_3$$

$$m_A(x) = 0, x > a_3$$

$$m_A(x) = 0, x < a_1$$

$$m_A(x) = \frac{x - a_1}{a_2 - a_1}, a_1 \leq x \leq a_2$$

$$m_A(x) = 1, a_2 \leq x \leq a_3$$

$$m_A(x) = \frac{a_1 - x}{a_1 - a_3}, a_3 \leq x \leq a_1$$

$$m_A(x) = 0, x > a_1$$

Where x is the time value for each layer and α_1, α_2 and α_3 are the lower middle and upper bound values of the fuzzy sets for time classification. The degree of membership of x to different fuzzy sets are then determined. Based on membership values, a process for determining precise actions to be applied must be developed. This task involves writing a rule set that provides an action for a given x. The formation of the rule set is comparable to that of an expert system, except that the rules incorporate linguistic variables with which humans are comfortable. The use of fuzzy sets allows rules to be derived easily. Fuzzy If-Then rules then can be built. The fuzzy rules could be of the form:

If [a_1 is (A_1) and a_2 is (A_2) and ...] Then [b_1 is (B_1) ALSO b_2 is (B_2) ALSO....]

Where A_i is the fuzzy set characterizing the respective decision variables and B_i is the fuzzy set characterizing the action variables. Although all possible conditions in the physical system seem imposing at first, the incorporation of fuzzy terms into the rules makes the development much easier. The fuzzy rules (A, B) associate an output fuzzy set B of the action values with an input fuzzy set A of input-variable values. The fuzzy rules are then written as antecedent-consequent pairs of If-Then statements. For example:

IF *the success of operational layer is* Very High Then *the time consumption for tactical layer is* High

The overall fuzzy output is derived by applying the "max" operation to the qualified fuzzy outputs each of which is equal to the minimum of the firing strength and the output membership function for each rule. Various schemes have been proposed to choose the final crisp output based on the overall fuzzy output. In this research study a type of inference method called center of gravity is used and illustrated in Equation 1 and Figure 7.

$$Output = \frac{\sum_{i=1}^{n} \alpha_i \, \mu_i}{\alpha_i} \tag{1}$$

Where α_i the upper bound value of the fuzzy set *i* and μ_i mi is the membership value of the fuzzy set *i*.

'What-if' scenarios can be performed using FCM and the fuzzy logic system for IT management decision making with time delay consideration. Table 4 shows several simulations where different IT management concepts based on FCM, as shown in Figure 5 have been turned on while all others have remained off. The time delays are calculated and results are presented.

To illustrate the risk analysis and time delay in the FCM model for IT management we will begin by using the event occurs row number 1 in Table 4.

Resource planning + service planning + development planning increase the probability of success at the tactical level. This then carries through to the strategic

Table 4. Time delay and consequences of different scenarios based on different 'what if' simulations

	What if the event occurs	Consequences
1	C8, C7, C6	$C8, C7, C6 \xrightarrow{2.3} C5 \xrightarrow{.80} C1$ Time delay in tactical layer is very low and time delay in strategic layer is low
2	C15, C12, C11, C3	$C15, C12 \xrightarrow{1.70} C11 \xrightarrow{.90} C5, C3 \xrightarrow{1.50} C1$ Time delay in operational layer is low and time delay in tactical layer is very low and time delay strategic layer is very low
3	C4, C7, C9, C10	$C10, C9, C7 \xrightarrow{2.20} C5, C4 \xrightarrow{1.50} C1$ Time delay Time in operational layer is very low and time delay in tactical layer is very low and time delay strategic layer is very low
4	C2, C8, C12, C14	$C14, C12 \xrightarrow{1.60} C11, C8 \xrightarrow{1.70} C5, C2 \xrightarrow{1.40} C1$ Time delay in operational layer is very low and time delay in tactical layer is very low and time delay strategic layer is very low
5	C7, C8, C9, C15	$C15 \xrightarrow{.80} C11, C7, C8, C9 \xrightarrow{3.10} C5 \xrightarrow{.80} C1$ Time delay in operational layer is low and time delay in tactical layer is very low and time delay strategic layer is very low
6	C4, C10, C13, C15	$C15, C13 \xrightarrow{1.60} C11, C10 \xrightarrow{1.70} C5, C4 \xrightarrow{1.50} C1$ Time delay in operational layer is low and time delay in tactical layer is very low and time delay strategic layer is very low
7	C7, C9, C12, C13	$C13, C12 \xrightarrow{1.70} C11, C9, C7 \xrightarrow{2.30} C5 \xrightarrow{.80} C1$ Time delay in operational layer is low and time delay in tactical layer is very low and time delay strategic layer is very low

level increasing the possibility of success. Successful completion of these events reduces time delay in the tactical layer and in the strategic layer. Simulation results in Table 4 show the influences of different sub-processes as well as any time delay caused by these sub-processes.

SUMMARY AND FINDINGS

Due to rapid changes in emerging technologies there is a need for constant improvement and adjustment of IT processes in organizations. Success of an organization is heavily dependent on the impact of IT. CIOs need to continuously monitor and analyze the performance of IT processes. They need to consider large numbers of activities performed by each IT process in their organization. The interdependencies of sub-processes make it very difficult for CIOs to comprehend and be fully aware of the effect of inefficiencies in IT processes and skilled personnel to the whole of IT and the organization itself.

This research study considered three layers of IT management consisting of strategic, tactical, and operational layers. With the complexity of this model consisting of large numbers of interdependent sub-processes, managing and monitoring IT systems are becoming increasingly difficult, and as such many IT management developments and implementations may be flawed. IT management models do not provide any facilities to analyze and assess different risks that may exist in such models in a systematic way.

Fuzzy Cognitive Maps (FCM) are employed to provide facilities to capture and represent complex relationships in IT management models and their processes and to improve the understanding of CIOs to analyze risks. Using FCM, different scenarios are considered. The proposed FCM model is used in conjunction with the proposed Luftman IT management model (Luftman, Bullen et al., 2004) to provide CIOs with possibilities for 'what-if' analysis.

By using FCMs, CIOs can regularly review and improve their IT management and provide greater improvement in development, monitoring and maintenance of IT facilities (for example providing adequate hardware to handle existing and future applications, providing systems that are easy to learn and easy to use, providing systems that are easy to maintain and upgrade, providing adequate budgets for their IT process developments). The CIO can perform 'what-if' analysis to find out the answer to questions such as:

Are the right technologies being used and are they integrated properly?

What levels of information sharing, security, and access should be supported?

Which applications will be developed versus which will be bought?

Are applications, tools, and data easy to upgrade, update, and maintain?

Who will upgrade, update, and maintain the applications, tools, and data?

Who will assess whether the architecture will meet the firm's needs?

Will the horizontal architecture support the horizontal processes of the firm?

CIO's also require the knowledge about the possible amount of time delay caused by malfunctioning of a sub-process in different layers of IT management (Luftman, Bullen et al., 2004), as such time delays need to be considered in IT management and decision making. This research considered the application of fuzzy logic to represent time delays in IT management layers of Luftman's IT management model. Simulation results show that fuzzy logic can be applied to this IT management model to assist in predicting possible time delays caused by malfunctioning of a sub-process in different layers of IT management.

REFERENCES

Aguilar, J. (2005). A survey about fuzzy cognitive maps papers. *International Journal of Computational Cognition, 3*(2), 27–33.

Andreou, A. S., Mateou, N. H., & Zombanakis, G. A. (2003). Evolutionary fuzzy cognitive maps: A hybrid system for crisis management and political decision making. *Proceedings of the International Conference on Computational Intelligence for Modelling, Control and Automation*, 2003.

Axelrod, R. (1976). Structure of decision. In *The cognitive maps of political elite* (1st ed.). Princeton University Press.

Carlsson, C., & Fuller, R. (1956). Adaptive fuzzy cognitive maps for hyper knowledge representation in strategic formation process. *Proceedings of the International Panel Conference on Soft and Intelligent Computing,* Budapest, 1996.

Georgopoulous, V. C., Malandrak, G. A., & Stylios, C. D. (2002). A fuzzy cognitive map approach to differential diagnosis of specific language impairment. *Artificial Intelligence in Medicine, 29*(3), 1–18.

Kosko, B. (1986). Fuzzy cognitive maps. *International Journal of Man-Machine Studies, 24*, 65–75. doi:10.1016/S0020-7373(86)80040-2

Kosko, B. (1988). Hidden patterns in combined and adaptive knowledge networks. *International Journal of Approximate Reasoning, 2*(4). doi:10.1016/0888-613X(88)90111-9

Kosko, B. (1997). *Fuzzy engineering*. Upper Saddle River, NJ: Prentice Hall.

Luftman, J., Bullen, V. C., Liao, D., Nash, E., & Neumann, C. (2004). *Managing the Information Technology resources – Leadership in information age*. Pearson Prentice Hall.

Salmeron, J. L. (2009). Supporting decision makers with fuzzy cognitive maps. *Research Technology Management, 52*(3), 53-59. Retrieved from http://web.eb-scohost.com

Taber, W. R. (1991). Knowledge processing with fuzzy cognitive maps. *Expert Systems with Applications, 2*(1). doi:10.1016/0957-4174(91)90136-3

Tsadiras, A. K., Kouskouvelis, I., & Margaritis, K. G. (2001). Making political decision using fuzzy cognitive maps: The FYROM crisis. *Proceedings of the 8th Panhellenic Conference on Informatics*, 2001.

Turban, E., Sharda, R., & Delen, D. (2011). *Decision support and business intelligence systems* (9th ed.). Upper Saddle River, NJ: Prentice Hall.

Van Gelder, T. J. (2010). The wise delinquency of decision makers. *Quadrant, 54*(3), 40-43. Retrieved from http://search.informit.com.au

Zadeh, L. A. (1965). Fuzzy sets. *Information and Control, 8*, 338–352. doi:10.1016/S0019-9958(65)90241-X

APPENDIX A

Fuzzy Rules Created for Time Delay IT Management Model

IF *the success of operational layer is* Very High Then *the time consumption for tactical layer is* Very High

IF *the success of operational layer is* High Then *the time consumption for tactical layer is* High

IF *the success of operational layer is* Moderate Then *the time consumption for tactical layer is* High

IF *the success of operational layer is* Low Then *the time consumption for tactical layer is* Low

IF *the success of operational layer is* Very Low Then *the time consumption for tactical layer is* Low

IF *the success of tactical layer is* Very High Then *the time consumption for tactical layer is* Very High

IF *the success of tactical layer is* High Then *the time consumption for tactical layer is* High

IF *the success of tactical layer is* Moderate Then *the time consumption for tactical layer is* Moderate

IF *the success of tactical layer is* Low Then *the time consumption for tactical layer is* Very Low

IF *the success of tactical layer is* Very Low Then *the time consumption for tactical layer is* Very Low

IF *the success of strategic layer is* Very High Then *the time consumption for tactical layer is* Very High

IF *the success of strategic layer is* High Then *the time consumption for tactical layer is* High

IF *the success of strategic layer is* Moderate Then *the time consumption for tactical layer is* Moderate

IF *the success of strategic layer is* Low Then *the time consumption for tactical layer is* Very Low

Compilation of References

Adeyeye, M. (2008). E-commerce, business methods and evaluation of payment methods in Nigeria. *Electronic Journal of Information System Evaluation, 11*(1), 1–6.

Aguilar, J. (2005). A survey about fuzzy cognitive maps papers. *International Journal of Computational Cognition, 3*(2), 27–33.

AIT. (2001). *Case study of Key Corp (NYSE: KEY).*

AIT. (2001). *Case study of the Lloyds TSB Group.*

AIT. (2001). *Case study of the Woolwich.*

Al-Mashari, M. (2002). Electronic commerce: A comparative study of organizational experiences. *Benchmarking: An International Journal, 9*(2), 182–189. doi:10.1108/14635770210421845

Amazon. (n.d.). *Website.* Retrieved from www.Amazon.com

Andreou, A. S., Mateou, N. H., & Zombanakis, G. A. (2003). Evolutionary fuzzy cognitive maps: A hybrid system for crisis management and political decision making. *Proceedings of the International Conference on Computational Intelligence for Modelling, Control and Automation, 2003.*

Anumba, C. J., Baugh, C., & Khalfan, M. A. (2002). Organsiational structures to support concurrent engineering in construction. *Industrial Management & Data Systems, 102*(5), 260–270. doi:10.1108/02635570210428294

Arendt, C., Landis, R., & Meister, T. (1995). *The human side of change – Part 4* (pp. 22–27). IIE Solutions.

Avgerou, C., & Cornford, T. (1998). *Developing Information Systems: Concepts, issues and practice.* Palgrave Publisher.

Awallmah,N. A. H.(2002) Electronic government and the future of management: an empirical study in the public sector in Qatar country, magazine studies.

Axelrod, R. (1976). Structure of decision. In *The cognitive maps of political elite* (1st ed.). Princeton University Press.

Baets, W. (1996). Some empirical evidence on IS strategy alignment in banking. *Information & Management, 30*(4), 155–177. doi:10.1016/0378-7206(95)00056-9

Baiden, B. K., Price, A. F., & Dainty, A. R. (2003). Looking beyond processess: Human factors in team integration. In Greenwood, D. J. (Ed.), *ARCOM*. Brighton.

Baldrige National Quality Program. (2001). *TriView national bank case study*. Case study of TriView.

Baldwin, A. N., Thorpe, A., & Carter, C. (1999). The use of electronic information exchange on construction alliance projects. *Automation in Construction, 8*, 651–662. doi:10.1016/S0926-5805(98)00110-1

Barbour. (2002). *The Barbour Report 2002: Exploring the Web as an information tool: A practical guide/* Windsor, Canada: Barbour Index.

Barbour. (2003). *The Barbour Report 2003: Influencing clients: The importance of the client in product selection*. Windsor, Canada: Barbour Index.

Bharadwaj, S. (2000). A resource-based perspective on information technology capability and firm performance: an empirical investigation. *Management Information Systems Quarterly, 24*(1), 169–197. doi:10.2307/3250983

Boateng, R., Hinson, R. E., Heeks, R., & Molla, A. (2008). E-commerce in least developing countries: Summary evidence and implications. *Journal of African Business, 9*(2). doi:10.1080/15228910802479919

Boateng, R., Heeks, R., Molla, A., & Hinson, R. (2009). E-commerce in Ghana: Where we are and where we are headed. In Hinson, R., Boateng, R., & Mbarika, V. A. (Eds.), *E-commerce and customer management in Ghana* (pp. 23–59).

Booty, F. (1998). Network management: The bottom line. *Manufacturing Computer Solutions, 4*(5), 37–40.

Bowonder, B., Akshay, J., & Narendra, K. (2000). *E-governance in a fisherman's community: A case study of Pondicherry*. Case Study of Pondicherry.

Brandon, D. (2006). *Project management for modern Information Systems*. USA: IRM Press.

Broadbent, M., Weill, P., & St Clair, D. (1999). The implications of Information Technology infrastructure for business process redesign. *Management Information Systems Quarterly, 23*(2), 159–182. doi:10.2307/249750

Bromley, S., Worthington, J., & Robinson, C. (2003). *The impact of integrated teams on the design process*. London, UK: Construction Productivity Network.

Bruque, S., & Moyano, J. (2007). Organizational determinants of information technology adoption and implementation in SMEs: The case of family and cooperative firms. *Technovation, 27*(5), 241–253. doi:10.1016/j.technovation.2006.12.003

Compilation of References

Bui, T. X., Sankaran, S., & Sebastian, I. M. (2003). A framework for measuring national ereadiness. *International Journal of Electronic Business*, *1*(1), 3–22. doi:10.1504/IJEB.2003.002162

Carlsson, C., & Fuller, R. (1956). Adaptive fuzzy cognitive maps for hyper knowledge representation in strategic formation process. *Proceedings of the International Panel Conference on Soft and Intelligent Computing*, Budapest, 1996.

Caroline, C., & Paul, M. (2000). *From EDI to Internet commerce; the BHP Steel experience*. Retrieved from www.emerald-library.com

Carter, C., White, E., Hassan, T., Shelbourn, M., & Baldwin, A. (2002). Legal issues of collaborative electronic working in construction. *Proceedings of the Institution of Civil Engineers, Civil Engineering, Special Issue Two: Information Technology - The Key to Collaboration*, (pp. 10-16).

Cash, J. I., McFarlan, W. F., McKenney, J. L., & Applegate, L. M. (1992). *Corporate Information Systems management: Text and cases* (3rd ed.). Homewood, IL: Irwin.

Chan, S. (1997). In search of Democratic peace: Problems and promise. *Mershon International Studies Review*, *41*, 59–91. doi:10.2307/222803

Chanopas, A., Krairit, D., & Khang, D. B. (2006). Managing information technology infrastructure: A new flexibility framework. [Emerald Group Publishing Limited.]. *Management Research News*, *29*(10), 632–651. doi:10.1108/01409170610712335

Checkland, P., & Holwell, S. (1998). *Information, systems and information systems*. Chichester, UK: John Wiley.

Choucri, N., Maugis, V., Madnick, S., & Siegel, M. (2003). *Global e-readiness - for what?* MIT Sloan School of Management.

Cisco. (2001). *Cisco and Blue Nile*. Retrieved from http:// www.cisco.com/warp/ public1779/ ibs/solutions/ ecommercelbluenile cp.pdf

Clarke, S., & Doherty, N. (2004). The importance of a strong business-IT relationship for the realisation of benefits in e-business projects: The experiences of Egg. *Qualitative Market Research; Bradford*, *7*(1), 58-66.

Claudia, L., & Stefan, S. (2001). *Web portfolio based electronic commerce: The case of Transtec AG*. Retrieved from www.emerald-library.com/ft

Clegg, C., Axtell, C., Damodaran, L., Farbey, B., Hull, R., & Lloyd-Jones, R. (1997). Information Technology: A study of performance and the role of human and organizational factors. *Ergonomics*, *40*(9), 851–871. doi:10.1080/001401397187694

Cornick, T., & Mather, J. (1999). *Construction project teams: Making them work profitably*. London, UK: Thomas Telford Department for Environment, Transport and the Regions. (2000). *Construction statistics annual*. London.

Dada, D. (2006). E-readiness for developing countries: moving the focus from the environment to the users. *Electronic Journal of Information in Developing Countries, 27*(6), 1–14.

Damanpour, F. (1991). Organizational innovation: A meta-analysis of effects of determinants and moderators. *Academy of Management Journal, 34*(3), 555–590. doi:10.2307/256406

David, S. (2001). *eBay creates technology architecture for the future*. Sun Microsystems.

Dehning, B., Richardson, V., & Stratopoulos, T. (2005). Information technology investments and firm value. *Information & Management, 42*(7), 989–1008. doi:10.1016/j.im.2004.11.003

Dell. (2005, 2006, 2007 & 2008). *Customer case studies*. Dell. Retrieved from http:// www.dell.com/content/topics/ global.aspx/casestudies/

Department of Trade and Industry. (2004). *Business in the information age: The international benchmarking study*. London, UK: Author.

Dewett, T., & Jones, G. R. (2001). The role of information technology in the organization: A review, model and assessment. *Journal of Management, 27*(3), 313–346.

E-Business W@tch. (2006). *ICT and e-business in the construction industry*. Sector Report No. 7, European Commission.

EDS. (2007 & 2008). *Case studies*. EDS. Retrieved from http://www.eds.com/services/casestudies/

Effah, J., & Light, B. (2009a). *Beyond the traditional 'SME challenges' discourse: A historical field study of a dot.com failure in Ghana*. Paper presented at the UK Academy for Information System (UKAIS) 14th Annual Conference.

Effah, J., & Light, B. (2009b). *Understanding SME E-business challenges in developing countries: The case of dot.coms in Ghana*. Paper presented at the British Academy of Management (BAM) Conference.

Erdogan, B., Anumba, C., Bouchlaghem, D., & Nielsen, Y. (2010). An innovative integrated framework towards effective collaboration environments in construction. *International Journal of Technology Management, 50*(2), 139–168. doi:10.1504/IJTM.2010.032270

Farabi case. (2002). *Farabi connects remote customers and branch offices to corporate host for Caterpillar equipment dealer*. Case study of Al-Bahar.

FedEx. (n.d.). *Home page*. Retrieved from www.FedEx.com

Feeny, D., & Willcocks, L. (1998). Core IS capabilities for exploiting Information Technology. *Sloan Management Review*, 9–21.

FNS Case. (2001). *How a new approach to IT turned around the fortunes of Taiwan's Chinatrust Commercial Bank.* Case study of Chinatrust Commercial Bank.

Foster, W., Goodman, S., Osiakwan, E., & Bernstein, A. (2004). Global diffusion of the Internet IV: The Internet in Ghana. *Communications of the Association for Information Systems, 13*(1), 654–681.

Compilation of References

Fotr, J. (2006). *Managerial decision making*. Prague, Czech Republic: Ekopress.

Frempon, G., Essegbey, G., & Tetteh, E. O. (2007). *Survey on the use of mobile phones for micro and smal business development: the case of Ghana*. Accra, Ghana: CSIR-Science and Technology Policy Research.

Fuchs, C., & Horak, E. (2008). Africa and the digital divide. *Telematics and Informatics, 25*, 99–116. doi:10.1016/j.tele.2006.06.004

Fujitsu. (2000). *Case studies*. Fujitsu. Retrieved from http://www.fujitsu.com/global/casestudies/

Gartner Group. (2001). *Home page*. Retrieved from http://www.gartner.com/technology/home.jsp

Georgopoulous, V. C., Malandrak, G. A., & Stylios, C. D. (2002). A fuzzy cognitive map approach to differential diagnosis of specific language impairment. *Artificial Intelligence in Medicine, 29*(3), 1–18.

Grasseova, M., Dubec, R., & Horak, R. (2008). *Procedural management in public and private sectors*. Brno, Czech Republic: Computer Press.

Grasseova, M., Dubec, R., & Rehak, D. (2010). *The analysis of enterprise on manager's hands: 33 the most frequently applied methods of strategic management*. Brno, Czech Republic: Computer Press.

Grasseova, M. (2006). The implementation of SWOT analysis in long-term planning. *Defence and Strategy, 2*(6), 48-55. ISSN 1214-6463

Grover, G. (1998) *Identification of factors affecting the implementation of data warehousing*. Unpublished PhD Dissertation. Auburn University at Montgomery.

Hammer, M., & Champy, J. (1993). *Reengineering the corporation: A manifesto for the business revolution*. New York, NY: Harper Business.

Hammer, M., & Stanton, S. (1995). *The re-engineering revolution*. New York, NY: Harper Collins.

Harrington, H. J. (1991). *Business process improvement*. US: McGraw-Hill.

Harris, P. R., & Harris, K. G. (1996). Managing effectively through teams. *Team Performance Management, 2*(3), 22–36. doi:10.1108/13527599610126247

Hassan, T., Shelbourn, M., & Carter, C. (2008). *Collaboration in construction: Legal and contractual issues in ICT application*.

Hegazy, T., Zaneldin, E., & Grierson, D. (2001). Improving design coordination for building projects. I: Information model. *Journal of Construction Engineering and Management, 127*(4), 322–329. doi:10.1061/(ASCE)0733-9364(2001)127:4(322)

Hinson, R., & Boateng, R. (2007). Perceived benefits and management commitment to e-business usage in selected Ghanaian tourism firms. *Electronic Journal of Information Systems in Developing Countries, 31*(50), 1–18.

Holt, G. D., Love, P. E. D., & Nesan, L. J. (2000). Employee empowerment in construction: An implementation model for process improvement. *Team Performance Management, 6*(3/4), 47–51. doi:10.1108/13527590010343007

Hwang, H. G., Ku, C. Y., Yen, D. C., & Cheng, C. C. (2004). Critical factors influencing the adoption of data warehouse technology: A study of the banking industry. *Decision Support Systems, 37*, 1–21. doi:10.1016/S0167-9236(02)00191-4

IBM. (2001, 2006, 2007 & 2008). *Case studies*. IBM. Retrieved from http://www-01.ibm.com/software/success/

Ifinedo, P. (2005). Measuring Africa's e-readiness in the global networked economy: A nine-country data analysis. *International Journal of Education and Development using Information and Communication Technology, 1*(1), 53-71.

International Telecommunication Union. (2009). *Internet usage statistics for Africa.* Retrieved January 24, 2009, from http:// www.internetworldstats.com/ stats1.htm#africa

Jerman-Blažič, B. (2008). Web-hosting market development status and its value as an indicator of a country's e-readiness. *Telecommunications Policy, 32*, 422–435. doi:10.1016/j.telpol.2008.04.007

Kapurubandara, M., & Lawson, R. (2008). Availability of e-commerce support for SMEs in developing countries. *The International Journal on Advances in ICT for Emerging Regions, 1*(01), 3–11.

Kayworth, T., & Sambamurthy, V. (2000). Managing the information technology infrastructure. *Baylor Business Review, 18*(1), 13–15.

Keil, M., Mann, J., & Rai, A. (2000). Why software projects escalate: An empirical analysis and test of four theoretical models. *Management Information Systems Quarterly, 24*(4), 631–664. doi:10.2307/3250950

Kim, B. G., & Lee, S. (2007). Factors affecting the implementation of electronic data interchange in Korea. *Computers in Human Behavior, 24*(2), 263–283. doi:10.1016/j.chb.2006.11.002

Kitchen, P. J., & Daly, F. (2002). Internal communication during change management. *Corporate Communications, 7*(1), 46–53. doi:10.1108/13563280210416035

Klein, H. K., & Myers, M. D. (1999). A set of principles for conducting and evaluating interpretive field studies in Information Systems. *Management Information Systems Quarterly, 23*(1), 67–93. doi:10.2307/249410

Kosko, B. (1986). Fuzzy cognitive maps. *International Journal of Man-Machine Studies, 24*, 65–75. doi:10.1016/S0020-7373(86)80040-2

Kosko, B. (1988). Hidden patterns in combined and adaptive knowledge networks. *International Journal of Approximate Reasoning, 2*(4). doi:10.1016/0888-613X(88)90111-9

Kosko, B. (1997). *Fuzzy engineering*. Upper Saddle River, NJ: Prentice Hall.

Compilation of References

Kumar, R. (2002). Managing risks in IT projects: An options perspective. *Information & Management, 40*, 63–74. doi:10.1016/S0378-7206(01)00133-1

Kwon, T. H., & Zmud, R. W. (1987). Unifying the fragmented models of information systems implementation. In Borland, R. J., & Hirschhiem, R. A. (Eds.), *Critical issues in information systems research* (pp. 252–257). New York, NY: John Wiley.

Lauria, E., & Duchessi, P. (2007). A methodology for developing Bayesian networks: An application to information technology (IT) implementation. *European Journal of Operational Research, 179*(1). Elsevier Science.

Luftman, J., Bullen, V. C., Liao, D., Nash, E., & Neumann, C. (2004). *Managing the Information Technology resources – Leadership in information age.* Pearson Prentice Hall.

Lyytinen, K., & Hirschheim, R. (1987). Information failures—A survey and classification of the empirical literature. *Oxford Surveys in Information Technology, 4*, 257–309.

Mak, S. (2001). A model of information management for construction using information technology. *Automation in Construction, 10*, 257–263. doi:10.1016/S0926-5805(99)00035-7

Mbarika, V. W. A., Okoli, C., Byrd, T. A., & Datta, P. (2005). The neglected continent of IS research: A research agenda for Sub-Saharan Africa. *Journal of the Association for Information Systems, 6*(5), 130–170.

McAdam, R., & Galloway, A. (2005). Enterprise resource planning and organisational innovation: A management perspective. *Industrial Management & Data Systems, 105*(3), 280–290. doi:10.1108/02635570510590110

McFadden, F. R. (1996). Data warehouse for EIS: Some issues and impacts. *Proc. of the 29th Hawaii International Conference on System Sciences* (pp. 120-129).

Microsoft. (2007 & 2008). *Case studies.* Retrieved from http://www.microsoft.com/casestudies/

Milis, K., & Mercken, R. (2004). The use of the balanced scorecard for the evaluation of Information and Communication Technology projects. *International Journal of Project Management, 22*, 87–97. doi:10.1016/S0263-7863(03)00060-7

Molla, A., & Licker, P. S. (2005). Perceived e-readiness factors in e-commerce adoption: An empirical investigation in a developing country. *International Journal of Electronic Commerce, 10*(1), 83–110.

Motawa, I. A., Anumba, C. J., Lee, S., & Pena-Mora, F. (2007). An integrated system for change management in construction. *Automation in Construction, 16*, 368–377. doi:10.1016/j.autcon.2006.07.005

Motwani, J., Subramanian, R., & Gopalakrishna, P. (2005). Critical factors for successful ERP implementation: Exploratory findings from four case studies. *Computers in Industry, 56*, 529–544. doi:10.1016/j.compind.2005.02.005

Mumford, E. (1995). Creative chaos or constructive change: Business process-re-engineering versus socio-technical design. In Burke, G., & Peppard, J. (Eds.), *Examining business process re-engineering: Current perspectives and research directions* (pp. 192–216). New York, NY: Kogan Page.

Murdoch and Hughes. (2008). *Construction contracts: Law and management*. Spon Press.

Murray, P., & Donegan, K. (2003). Empirical linkages between firm competencies and organisational learning. *The Learning Organization, 10*(1), 51. doi:10.1108/09696470310457496

O'Brien, M. J., & Al-Biqami, N. M. (1999). Information Technology and the structure of the Saudi Arabian construction industry. *8th International Conference on Durability of Building Materials and Components*, (vol. 4, pp. 2327-2337). Ottawa, Canada: NRC Research Press, National Research Council of Canada.

Osei-Brysona, K. M., & Kob, M. (2004). Exploring the relationship between information technology investments and firm performance using regression splines analysis. *Information & Management, 42*, 1–13. doi:10.1016/j.im.2003.09.002

Paulson, B. (1995). *Computer applications in construction*. Singapore: McGraw-Hill Inc.

Pena-Mora, F., & Dwivedi, G. H. (2002). Multiple device collaborative and real time analysis system for process management in civil engineering. *Journal of Computing in Civil Engineering, 16*(1), 23–37. doi:10.1061/(ASCE)0887-3801(2002)16:1(23)

PIECC. (2006). Planning and implementation of effective collaboration within construction/ Retrieved July 10, 2009, from http://piecc.lboro.ac.uk/

Poon, P., & Wagner, C. (2001). Critical success factors revisited: Success and failure cases of Information Systems for senior executives. *Decision Support Systems, 30*(3), 393–418. doi:10.1016/S0167-9236(00)00069-5

Premkumar, G., & Ramamurthy, K. (1995). The role of inter-organizational and organizational factors in the decision mode for adoption of inter-organizational systems. *Decision Sciences, 26*(3), 303–336. doi:10.1111/j.1540-5915.1995.tb01431.x

Premkumar, G., & Roberts, M. (1999). Adoption of new information technologies in rural small businesses. *Omega International Journal of Management Science, 27*(4), 467–484. doi:10.1016/S0305-0483(98)00071-1

Proctor, T., & Doukakis, I. (2003). Change management: The role of internal communication and employee development. *Corporate Communications: An International Journal, 8*(4), 268–277. doi:10.1108/13563280310506430

Purcell, F., & Toland, J. (2004). Electronic commerce for the South Pacific: A review of e-readiness. *Electronic Commerce Research, 4*, 241–262. doi:10.1023/B:ELEC.0000027982.96505.c6

Compilation of References

Ragu-Nathan, B. S., Apigian, C. H., Ragu-Nathan, T. S., & Tu, Q. (2004). A path analytic study of the effect of top management support for information systems performance. *Omega, 32,* 459–471. doi:10.1016/j.omega.2004.03.001

Rajiv, D., & Robert, J. (1998). Case study of electronic banking at Meridian Bancorp. *Information and Software Technology, 33*(1), 1–9.

Ranganathan, C., & Kannabiran, G. (2004). Effective management of information systems function: An exploratory study of Indian organizations. *International Journal of Information Management, 24,* 247–266. doi:10.1016/j.ijinfomgt.2004.02.005

Ratnasingam, P. (2001). Inter-organizational trust in EDI adoption: The case of Ford Motor Company and PBR Limited in Australia. *Internet Research; Bradford, 11*(3), 261-268.

Rehak, D., & Dvorak, J. (2010). Risk catalogue as a software tool for supporting the business continuity planning. *Int. J. Business Continuity and Risk Management, 1*(2), 187-196. ISSN 1758-2164

Rehak, D., Dubec, R., & Grasseova, M. (2008). *Assessment of external environmental elements within SWOT analysis utilizing the evaluation of risks and benefits.* Paper presented at the Symposium on Risk Analysis /Management Cybernetics /Economics (19th International Conference on Systems Research Informatics & Cybernetics), Baden-Baden, Germany.

Rehak, D., Dvorak, J., & Grasseova, M. (2009). Principles, framework and process of risk management. In J. Navrátil & J. Barta (Ed.), *International Conference Security Management and Society* (pp. 364-376). Brno, Czech Republic: University of Defence. ISBN 978-80-7231-653-3

Rezgui, Y. (2007). Exploring virtual team-working effectiveness in the construction sector. *Interacting with Computers, 19*(1), 96–112. doi:10.1016/j.intcom.2006.07.002

Rezgui, Y., Wilson, I., Olphert, W., & Damodaran, L. (2005). Socio-organizational issues. In Camarinha-Matos, L. M., Afsarmanesh, H., & Ollus, M. (Eds.), *Virtual organizations systems and practices.* New York, NY: Springer. doi:10.1007/0-387-23757-7_13

Robson, W. (1997). *Strategic management and Information Systems.* London, UK: Pitman Publishing.

Rodriguez-Abitia, G., Vidrio, S., & Montiel-Sanchez, C. (2004). *Assessing the state of e-readiness for small and medium companies in Mexico: A proposed taxonomy and adoption model.* Paper presented at the AMCIS 2004 Proceedings. Paper 78.

Rodriguez-Repiso, L., Setchi, R., & Salmeron, J. (2007). Modelling IT projects success with Fuzzy Cognitive Maps. *Expert Systems with Applications, 32,* 543–559. doi:10.1016/j.eswa.2006.01.032

Rogers, E. (1995). *Diffusion of innovations* (4th ed.). New York, NY: Free Press.

Roshani, D., & Tizani, W. (2005). *Integrated IFC based collaborative building design using internet technology.* The Tenth International Conference on Civil, Structural and Environmental Engineering Computing, Rome, Italy.

Ruikar, K., Anumba, C. J., & Carillo, P. M. (2005). End-user perspectives on use of project extranets in construction organsiations. *Engineering, Construction, and Architectural Management, 12*(3), 222–235. doi:10.1108/09699980510600099

Rye, C. (1996). *Change management action kit.* London, UK: Kogan Page.

Saffu, K., Walker, J., & Hinson, R. (2008). Strategic value and electronic commerce adoption among small and medium-sized enterprises in a transitional economy. *Journal of Business and Industrial Marketing, 23*(6), 395–404. doi:10.1108/08858620810894445

Saleh, Y., & Alshawi, M. (2005). An alternative model for measuring the success of IS projects: The GPIS model. *The Journal of Enterprise Information Management, 18*(1), 47–63. doi:10.1108/17410390510571484

Salmeron, J. L. (2009). Supporting decision makers with fuzzy cognitive maps. *Research Technology Management, 52*(3), 53-59. Retrieved from http://web.ebscohost.com

Sauer, C., & Cuthbertson, C. (2003). *The state of IT project management in the UK.* Oxford: Templeton College.

Shah, M. H., & Siddiqui, F. A. (2006). Organizational critical success factors in adoption of e-banking at the Woolwich bank. *International Journal of Information Management, 26,* 442–456. doi:10.1016/j.ijinfomgt.2006.08.003

Singh, M., & Byrne, J. (2005). Performance evaluation of e-business in Australia. *Electronic Journal of Information Systems Evaluation, 8*(1), 71–80.

Slevin, D. P., & Pinto, J. K. (1986). The project implementation profile: New tool for project managers. *Project Management Journal, 17*(4), 57–70.

Small Business Service. (2004). *Statistics.* Retreived April 2, 2009, from http:// www.sbs.gov.uk/ default.php? page=/statistics/dcfault.php

Soliman, K. S., & Janz, B. D. (2004). An exploratory study to identify the critical factors affecting the decision to establish Internet-based interorganizational information systems. *Information & Management, 41,* 697–706. doi:10.1016/j.im.2003.06.001

Swanson, E. B., & Wang, P. (2005). Knowing why and how to innovate with packaged business software. *Journal of Information Technology, 20,* 20–31. doi:10.1057/palgrave.jit.2000033

Taber, W. R. (1991). Knowledge processing with fuzzy cognitive maps. *Expert Systems with Applications, 2*(1). doi:10.1016/0957-4174(91)90136-3

Tallon, P. P., Kraemer, K. L., & Gurbaxani, V. (2000). Executives' perceptions of the business value of information technology - A process-oriented approach. *Journal of Management Information Systems, 16*(4), 145–173.

Compilation of References

Tarafdar, M., & Vaidya, S. (2006). Challenges in the adoption of e-commerce technologies in India: The role of organizational factors. *International Journal of Information Management, 26*, 428–441. doi:10.1016/j.ijinfomgt.2006.08.001

Teo, T. S. H., Ang, J., & Pavri, F. (1997). The state of strategic IS planning practices in Singapore. *Information & Management, 33*, 13–23. doi:10.1016/S0378-7206(97)00033-5

The Bridges Organization. (2001). *Comparison of e-readiness assessment models.*

Thong, J. Y. L. (1999). An integrated model of information system adoption in small business. *Journal of Management Information Systems, 15*(4), 187–214.

Thorpe, T., & Mead, S. (2001). Project-specific web sites: Friend or foe? *Journal of Construction Engineering and Management, 127*(5), 406–413. doi:10.1061/(ASCE)0733-9364(2001)127:5(406)

Tomecek, P. (2008). The methods of computer attacks in the information warfare. In P. Hruza & P. Tomecek (Ed.), *The 2nd International Conference on Advanced and Systematic Research.* (pp. 79-82). Zagreb, Croatia: Faculty of Teacher Education of the University of Zagreb. ISBN 978-953-7210-14-4

Tsadiras, A. K., Kouskouvelis, I., & Margaritis, K. G. (2001). Making political decision using fuzzy cognitive maps: The FYROM crisis. *Proceedings of the 8th Panhellenic Conference on Informatics, 2001.*

Tucker, S. N., & Mohamed, S. (1996). Introducing information technology in construction: Pains and gains. *Proceedings of the CIB-W65 Symposium on Organization and Management of Construction,* Glasgow, (pp. 348-356).

Turban, E., Lee, J., King, D., & Shung, H. M. (2000). *Electronic commerce, a managerial perspective.* London, UK: Prentice Hall.

Turban, E., Sharda, R., & Delen, D. (2011). *Decision support and business intelligence systems* (9th ed.). Upper Saddle River, NJ: Prentice Hall.

Underwood, J., & Alshawi, M. (2000). Forecasting building element maintenance within an integrated construction environment. *Automation in Construction, 9*(2), 169–184. doi:10.1016/S0926-5805(99)00003-5

Vadapalli, A., & Mone, M. A. (2000). Information technology project outcomes: User participation structures and the impact of organization behavior and human resource management issues. *Journal of Engineering and Technology Management, 17*, 127–151. doi:10.1016/S0923-4748(00)00018-7

Van Gelder, T. J. (2010). The wise delinquency of decision makers. *Quadrant, 54*(3), 40-43. Retrieved from http://search.informit.com.au

Walsham, G. (1995). Interpretive case studies in IS research: nature and method. *European Journal of Information Systems, 4*, 74–81. doi:10.1057/ejis.1995.9

Walsham, G. (2006). Doing interpretive research. *European Journal of Information Systems*, *15*(3), 320–330. doi:10.1057/palgrave.ejis.3000589

Walsham, G., & Sahay, S. (1999). GIS for district-level administration in India: Problems and opportunities. *Management Information Systems Quarterly*, *23*(1), 39–66. doi:10.2307/249409

Weippert, A., & Kajewski, S. L. (2002). Internet-based information and communication systems on remote construction projects: A case study analysis. *Construction Innovation*, *2*(2), 103–116.

Wilkinson, P. (2005). *Construction collaboration technologies: The extranet evolution*. London, UK: Spon Press.

Willcocks, L., & Lester, S. (1994). Information Systems investments: Evaluations at the feasibility stage of project. *Technovation*, *11*(5), 283–302. doi:10.1016/0166-4972(91)90027-2

Wilson, R. (2006). *New digital state leaders emerge from e-government study*. Center for Digital Government. Retrieved from www.centerdigitalgov.com/ center/highlighstory.html

Zadeh, L. A. (1965). Fuzzy sets. *Information and Control*, *8*, 338–352. doi:10.1016/S0019-9958(65)90241-X

Zee, H. V. D. (2002). *Measuring the value of Information Technology*. Hershey, PA: Idea Group Publishing.

About the Contributors

Mustafa Alshawi is a Professor of IT in Construction at the University of Salford. He holds many international advisory posts in various countries and a consultant to the World Bank. He is also the Editor in Chief of the international journal *Construction Innovation: Information, Process, Management* and the author of more than hundred publications in fields such as integrated computer environments, databases, object oriented databases, IT strategies, CAD, planning automation, and IS success factors. Professor Alshawi is the founder of five innovative postgraduate courses ranging from PG Diploma, MSc, Master of Research, and PhD (without residence). His leading work in management of information in collaborative environments, integrated databases in construction and computer integration are internationally known. His computer environments, such as SPACE and WISPER, have set up a vision for future work practices, which enable different professionals to share and exchange project information.

Mohammed Arif is a Senior Lecturer in the school of Built Environment at the University of Salford. One of the major areas of research for Dr. Arif is IT management. He has published several articles in the areas of IT management, system selection, ERP, e-Government, knowledge management, and Malcolm Baldrige Quality Awards. He has an extensive track record of Masters and PhD supervision. He is holds several advisory positions and has acted as consultant to several corporations in the USA, Middle East, and the UK. Currently, two areas of IT management research that he is pursuing are usability of e-learning systems for different languages and virtual communities of practices.

* * *

Aisha Abuelmaatti earned her B.Sc. Multimedia and Internet Technology (2006) and M.Sc. Information Systems (2007) degrees from the University of Salford in UK, before carrying out her PhD in the School of the Built and Human Environment in the same university. Her PhD work is focused on Collaborative

Environments and its role in Architecture, Engineering, and Construction sectors. She conducts research in areas related to the applications of advanced informatics for collaborative working.

Vian Ahmed is currently the Director of Postgraduate Research Studies, at the School of Built Environment, University of Salford. She holds a BEng. in Civil Engineering, MSc in Construction and PhD in Computer Aided learning in Construction. Prof. Ahmed's research interest and expertise evolve around people, cultures and skills in construction, including the utilisation do IT, e-learning and project management skills in the industry. Prof. Ahmed has track record of over 70 conference publication, thirty refereed journal publications and a large pool of graduated PhD students. She had been the Director of the on-line PhD programme since 2005.

Ayman Abdualaziz Abdullah Altamem is Vice Dean for Development and Quality, College of Applied Studies and Community Service at University of Bradford, UK. He earned his PhD from University of Bradford in Information Technology, his MSc in Information Systems from London South Bank University, and his BSc in Information Systems from King Saud University, Saudi Arabia. He is a member of the ACM, BCS, and American Society for Information Science and Technology (ASIS&T).

John Effah completed his Ph.D. in Information Systems with the Informatics Research Institute (IRIS), University of Salford, UK. He is currently the Head of the Department of Operations & Management Information Systems (MIS), University of Ghana Business School. John holds a Bachelor of Science in Administration Degree in Accounting with first class honours and an MBA in MIS from the University of Ghana Business School. In the Business School, he teaches Information Systems Strategy at the Executive MBA level and E-Business at the MBA level. His key research interest borders on E-Business, E-Government and E-Health Systems. John had a considerable number of years experience in ICT consulting with Deloitte before joining the University of Ghana Business School as a lecturer. In this capacity, he managed a number of ICT projects in Ghana and abroad. He has also served as a reviewer for a number of international journals including, the European Journal of Information Systems, and the African Journal of Information Systems as well as serving on its editorial board. He has presented his research at various academic conferences such as the UK Academy of Information Systems (UKAIS), British Academy of Management with its E-Business Special Interest Group, and ICT for Africa. John is a member of the UK Academy of Information Systems (UKAIS), the E-Business Special Interest Group of the British Academy of Management, and the Association of Information Systems (AIS).

Monika Grasseova, PhD, was born in 1973. She graduated from the Faculty of Economics at the VSB-Technical University of Ostrava. In 2002 she acquired the academic degree of PhD in the branch called Economics and Management. Currently she works as a senior lecturer at the Faculty of Economics and Management of University of Defence in Brno. She specializes in the area of strategic management, process management, management by objectives, analytical tools of management and assessment methods. She teaches in the Course of General Staff, the Course of Senior Officers and other specialized courses. She is a guarantor and in charge of the courses of process management and audit for the Czech DoD personnel. She has also participated in the courses of process management, e.g. for the inspectors of the Supreme Audit Office of the Czech Republic. She is a research worker participating in projects and expertises within her specialization. She is an author and co-author of approximately 90 specialist publications, 25 of which have been published outside the Czech Republic.

Ben Light is Professor of Digital Media in the School of Media, Music and Performance, and a member of the Communication, Cultural and Media Studies Research Centre at the University of Salford, UK. His current research interests centre on analysing the development and use of the Internet in everyday life.

Eric Lou is a PhD student at the University of Salford. Eric's research interests encompass construction ICT, business strategies, e-readiness and corporate responsibility. His previous experience and work is related to the Construction Industry Development Board of Malaysia and the Public Works Department (developing National master plans and strategies). He has helped to develop Malaysia's National E-Tendering Initiative, and has been involved in procuring several other high level Reports e.g. 'Doing Construction Business' for Malaysian contractors, Reclassification of Malaysian Contractors, etc. Eric is now working in the area of social, economic and environment in Corporate Responsibility through integrated ICT.

David Rehak, PhD, was born in 1978. He graduated from the Military University of the Ground Forces in Vyskov, The Czech Republic, in the Economics of Environmental Protection. After that he started a 3-year doctoral study in the field of modelling and simulation of the protection of troops and inhabitants within the study programme "Protection of Troops and Inhabitants" which he finished in 2005 at the University of Defence in Brno. From 2006 to 2009 he worked as a senior lecturer at the University of Defence. Currently he works as a senior lecturer at the Faculty of Safety Engineering of VSB-Technical University of Ostrava. His scientific work is aimed at security environment, risk management, civil protection and environmental protection. He is an author and co-author of approx. 80 publications, 25 of which have been published outside the Czech Republic.

Yasser Saleh has over 20 years experience in academic teaching in the fields of Computer Science and Information Systems/Information Technology at Kuwait University, The Public Authority for Applied Education and Training, and Arab Open University. His doctoral work involved building a Readiness Model that enables evaluation of organisational readiness pre-implementation of Information Systems projects which includes studying Information Systems strategic planning and its alignment with organisational strategic planning. The thesis contributed to post-implementation evaluation theory of the success of Information Systems within organisations. His research interests include organizational readiness for implementation of Information Systems, success measurement post-implementation of Information Systems projects, e-learning, system development, and impact of social media on the formations of political systems. He has been involved in consultation work regarding Information Systems projects in private and public sectors in academic, banking, retail, oil, and non-profit organisations.

Hafez Salleh is a Deputy Dean(Undergraduate)/Senior Lecturer in the Faculty of Built Environment, University of Malaya, Malaysia. He obtained a BSc, MSc and PhD Degree from School of the Built Environment and top rated, 6* Research Institute for the Built and Human Environment, University of Salford, United Kingdom. He has 15 years experience in both academia and industry practice and has been an active researcher and research supervisor. He has supervised and supported a wide range of post graduate research students in various research themes. His main research interests are IT/IS performance evaluation and management, project management and sustainable design. His previous research was the development of IT/IS Readiness Model that focuses on evaluation of organizational readiness prior IT/IS investment. He is currently working on the theme of IT/IS evaluation in construction with a particular interest in the links between 'hard' and human issues. He is also the technical committee member of Malaysian Construction Industry Master Plan (CIMP) 2005-2015, Strategic Thrust 6: Leveraging Information Technology and Editorial Board Members, Journal of Surveying, Construction and Property.

Mohamed Zairi is the JURAN Chair in TQM and previously the SABIC Chair in Best Practice Management based at the European Centre for Best Practice Management, UK. Professor Zairi holds a BSc (Hon) in Polymer Sciences and Technology; MSc in Safety and Health and PhD in Management of Advanced Manufacturing Technology. He is currently The Executive Chairman of the European Centre for

Best Practice Management and Assistant Vice Chancellor for Strategy and Growth at Hamdan Bin Mohammed eUniversity in Dubai. Professor Zairi has been involved in guiding, mentoring and advising on the implementation of excellence in both the government and private sector context. He has also acted in the capacity of Jury Chairman of various prestigious international awards. Professor Zairi has been awarded The Ishikawa/Harrington Medal (2005) for his significant contribution in 12 Asian countries by the Asian Pacific Quality Organisation (APQO). He has also been awarded the Grand Master Six Sigma Medal (2005) for unique contribution to the growth of quality initiatives, the development of tools and systems and the impact of organisational business performance in various parts of the World. In 2007, he received the Lancaster Medal by The American Society for Quality (ASQ), for his outstanding contribution to contributions to the international fraternity of quality professionals. In 2010, he was presented with the 2009 ASQ Grant Medal for the development of exceptionally meritorious, technologically innovative, and intellectually challenging quality management educational programs. He was also presented with the Yoshio Kondo Academic Prize (2010) for outstanding research that has advanced the global body of quality knowledge.

Index

CPSIA information can be obtained at www.ICGtesting.com
Printed in the USA
BVOW060729141111

275933BV00001B/3/P